T0319395

Strategic Materials and Computational Design

Strategic Materials and Computational Design

A Collection of Papers Presented at the 34th International Conference on Advanced Ceramics and Composites January 24–29, 2010 Daytona Beach, Florida

Edited by
Waltraud M. Kriven
Yanchun Zhou
Miladin Radovic

Volume Editors
Sanjay Mathur
Tatsuki Ohji

A John Wiley & Sons, Inc., Publication

Published by John Wiley & Sons, Inc., Hoboken, New Jersey.
Published simultaneously in Canada.

For general information on our other products and services or for technical support, please contact our
Customer Care Department within the United States at (800) 762-2974, outside the United States at
(317) 572-3993 or fax (317) 572-4002.

Wiley also publishes its books in a variety of electronic formats. Some content that appears in print may
not be available in electronic format. For information about Wiley products, visit our web site at
www.wiley.com.

Library of Congress Cataloging-in-Publication Data is available.

ISBN 978-0-470-92191-3

Contents

COMPUTATIONAL DESIGN, MODELING, SIMULATION AND CHARACTERIZATION

NANOLAMINATED TERNARY CARBIDES AND NITRIDES

Preface

Contributions from three Focused Sessions that were part of the 34th International Conference on Advanced Ceramics and Composites (ICACC), in Daytona Beach, FL, January 24-29, 2010 are presented in this volume. The broad range of topics is captured by the Focused Session titles, which are listed as follows: FS1—Geopolymers and other Inorganic Polymers; FS3—Computational Design, Modeling Simulation and Characterization of Ceramics and Composites; and FS4—Nanolaminated Ternary Carbides and Nitrides (MAX Phases).

The session on Geopolymers and other Inorganic Polymers continues to attract growing attention from international researchers (USA, Australia, France, Germany, Italy, Czech Republic, and Viet Nam) and it is encouraging to see the variety of established and new applications being found for these novel and potentially useful materials. The session organizer gratefully acknowledges the support of the US Air Force Office of Scientific Research (AFOSR) through Dr. Joan Fuller. The AFOSR has continuously supported these conferences since the first meeting in Nashville, TN in 2003.

Focused Session 3 was dedicated to design, modeling, simulation and characterization of ceramics and composites. 27 technical papers were presented on prediction of crystal structure and phase stability, characterization of interfaces and grain boundaries at atomic scale, optimization of electrical, optical and mechanical properties, modeling of defects and related properties, design of materials and components at different length scales, application of novel computational methods for processing. Four of these papers are included in this issue of CESP.

Focused Session 4 was dedicated to MAX phases - a class of ternary carbides and nitrides with nanolaminated structure and general formula $M_{n+1}AX_n$ (where M is an early transition metal, A is an A-group element from IIIA to VIA, X is either C or N, and n=1, 2, 3 ...). The MAX phases have attracted recently a lot of attention because they possess unique combination of metallic- and ceramic-like properties. In all, 30 technical papers were presented during this session. Four of these papers are included in this issue.

WALTRAUD M. KRIVEN, University of Illinois at Urbana-Champaign Urbana, USA
YANCHUN ZHU, Institute of Metal Research, Chinese Academy of Sciences, China
MILADIN RADOVIC, Texas A&M University, USA

Introduction

This CESP issue represents papers that were submitted and approved for the proceedings of the 34th International Conference on Advanced Ceramics and Composites (ICACC), held January 24–29, 2010 in Daytona Beach, Florida. ICACC is the most prominent international meeting in the area of advanced structural, functional, and nanoscopic ceramics, composites, and other emerging ceramic materials and technologies. This prestigious conference has been organized by The American Ceramic Society's (ACerS) Engineering Ceramics Division (ECD) since 1977.

The conference was organized into the following symposia and focused sessions:

Symposium 1	Mechanical Behavior and Performance of Ceramics and Composites
Symposium 2	Advanced Ceramic Coatings for Structural, Environmental, and Functional Applications
Symposium 3	7th International Symposium on Solid Oxide Fuel Cells (SOFC): Materials, Science, and Technology
Symposium 4	Armor Ceramics
Symposium 5	Next Generation Bioceramics
Symposium 6	International Symposium on Ceramics for Electric Energy Generation, Storage, and Distribution
Symposium 7	4th International Symposium on Nanostructured Materials and Nanocomposites: Development and Applications
Symposium 8	4th International Symposium on Advanced Processing and Manufacturing Technologies (APMT) for Structural and Multifunctional Materials and Systems
Symposium 9	Porous Ceramics: Novel Developments and Applications
Symposium 10	Thermal Management Materials and Technologies
Symposium 11	Advanced Sensor Technology, Developments and Applications

Focused Session 1 Geopolymers and other Inorganic Polymers
Focused Session 2 Global Mineral Resources for Strategic and Emerging
                   Technologies
Focused Session 3 Computational Design, Modeling, Simulation and
                   Characterization of Ceramics and Composites
Focused Session 4 Nanolaminated Ternary Carbides and Nitrides (MAX Phases)

The conference proceedings are published into 9 issues of the 2010 Ceramic Engineering and Science Proceedings (CESP); Volume 31, Issues 2–10, 2010 as outlined below:

- Mechanical Properties and Performance of Engineering Ceramics and Composites V, CESP Volume 31, Issue 2 (includes papers from Symposium 1)
- Advanced Ceramic Coatings and Interfaces V, Volume 31, Issue 3 (includes papers from Symposium 2)
- Advances in Solid Oxide Fuel Cells VI, CESP Volume 31, Issue 4 (includes papers from Symposium 3)
- Advances in Ceramic Armor VI, CESP Volume 31, Issue 5 (includes papers from Symposium 4)
- Advances in Bioceramics and Porous Ceramics III, CESP Volume 31, Issue 6 (includes papers from Symposia 5 and 9)
- Nanostructured Materials and Nanotechnology IV, CESP Volume 31, Issue 7 (includes papers from Symposium 7)
- Advanced Processing and Manufacturing Technologies for Structural and Multifunctional Materials IV, CESP Volume 31, Issue 8 (includes papers from Symposium 8)
- Advanced Materials for Sustainable Developments, CESP Volume 31, Issue 9 (includes papers from Symposia 6, 10, and 11)
- Strategic Materials and Computational Design, CESP Volume 31, Issue 10 (includes papers from Focused Sessions 1, 3 and 4)

The organization of the Daytona Beach meeting and the publication of these proceedings were possible thanks to the professional staff of ACerS and the tireless dedication of many ECD members. We would especially like to express our sincere thanks to the symposia organizers, session chairs, presenters and conference attendees, for their efforts and enthusiastic participation in the vibrant and cutting-edge conference.

ACerS and the ECD invite you to attend the 35th International Conference on Advanced Ceramics and Composites (http://www.ceramics.org/icacc-11) January 23–28, 2011 in Daytona Beach, Florida.

Sanjay Mathur and Tatsuki Ohji, Volume Editors
July 2010

Geopolymers and Other Inorganic Polymers

GEOMATERIAL FOAM TO REINFORCE WOOD

E. Prud'homme, P. Michaud, C. Peyratout, A. Smith, and S. Rossignol
Groupe d'Etude des Matériaux Hétérogènes (GEMH-ENSCI)
47-73, avenue Albert Thomas, 87065 Limoges, France

E. Joussein
GRESE EA 4330, Université de Limoges
123, avenue Albert Thomas, 87060 Limoges, France

N. Sauvat
[3] Groupe d'Etude des Matériaux Hétérogènes (3MsGC-IUT)
Boulevard Jacques DERCHE, 19300 Egletons, France

ABSTRACT

The synthesis of geopolymers based on alkaline polysialate was achieved by the alkaline activation of raw minerals and silica fume. The in-situ inorganic foam is based on silica fume, an industrial waste, added to a solution containing dehydroxylated kaolinite and alkaline hydroxide pellets dissolved in potassium silicate. The foam was synthesized from the in-situ gaseous production of dihydrogen via oxidation of free silicon (present in the silica fume) by water in an alkaline medium. Considering its potential as an insulating material for applications in building geomaterials, this foam was used to make a wood-foam composite at room temperature. The composite's mechanical properties were studied in relation to its chemistry.

INTRODUCTION

Currently, there is a political as well as a societal demand for products which require less energy during the manufacturing process and are easy to recycle. New materials must exhibit analogous or improved properties with respect to those of existing materials.[1]

Materials associating mineral binders and local cellulose raw materials can be found in all cultures and time periods.[2] However, the physical chemistryof the exchanges between the various components of these composites is not well known. Furthermore, new cementitious materials known as geopolymers, including most silicates or aluminosilicates, can be used as substitutes for conventional hydraulic binders.[3, 4] Geopolymer materials may also be used to passivate industrial wastes. Unfortunately, control of consolidation time, hydraulic behaviour and mechanical properties of these materials remains difficult. In a previous paper, we demonstrated the synthesis of inorganic foam with insulating properties.[5]

In this work, we describe the preparation of various composite materials based on the junction of this foam and wood. First, we study the biocompatibility from a chemical basis between wood and mineral compound with pH-dependent decomposition analysis, thermal analysis, and scanning electron microscopy (SEM). We subsequently examine the mechanical properties of the composite with embedding tests to observe interfacial shear behavior.

EXPERIMENTAL PART

The initial geomaterial was prepared from a solution containing dehydroxylated kaolinite and KOH pellets (85.7% purity) dissolved in potassium silicate solution (Si/K = 1.70 by mole, density 1.20), as described in Figure 1a. An industrial waste such as silica fume (0.7%wt of free silicon, 97.5%wt of silica, 0.275%wt of carbon) was added to previous solution. The reaction mixture was then placed in a sealed polystyrene mold in an oven at 70°C for 6 h to complete the polycondensation reaction. The material was then removed from the mold and placed in an oven at 70°C for 24 h to dry. During this process, the in-situ inorganic foam based on silica fume was formed from the in-situ gaseous production of dihydrogen due to oxidation of free silicon (content in the silica fume) by water in alkaline medium, which was confirmed via TGA-MS experiments.

(a)

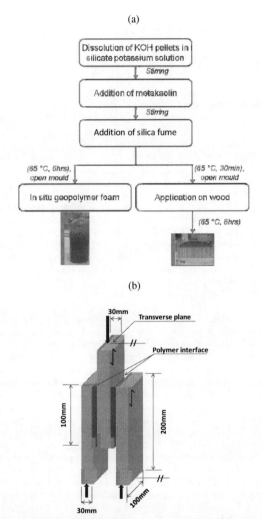

(b)

Figure 1: (a) Synthesis protocol of in-situ inorganic foam and wood/foam composites; (b) composite specimen for testing in double shear.

The durability of materials in contact with solution was investigated by immersing pieces in solution at various initial pH values. The solution's pH was monitored at 25°C with a MeterLab PHM240 at 25°C. To study wood, pieces of fir were immersed in potassium silicate solution, sodium hydroxide solution, potassium hydroxide solution and distilled water corresponding to initial pH value of 11.5, 12.6, 13.8 and 7.5, respectively. The ratio between the masses of wood and solution was 0.17 (with m_{wood}=8.3 g and $m_{solution}$=49 g). The pH values of solutions were then measured during the next 10 days, once a day. The foam was immersed in solutions prepared with hydrochloric acid, sodium hydroxide or distilled water with initial pH values of 2, 6.5 and 10.5, respectively. The pH values of solutions were subsequently measured over a period of 40 min, once every 2 min.

The preparation of the composite required joining together the foam and wood. The reactive mixture of the foam is relatively fluid, so it was necessary to activate the mixture in order to increase the viscosity, thereby facilitating the shaping of the material. The mixture was activated by placement in oven at 70°C for 30 min. Once made active, the mixture was deposited on wood kept at ambient temperature. In order to test the mechanical properties of the wood-foam interface and to compare it with the wood-industrial glue interface in double-shear, specimens were synthesized according to the model given in Figure 1b. Three pieces of wood (each 200x100x30 mm) were attached via two layers of foam (each 100x100x5 mm) or industrial glue. The thickness of each layer was set with a mold. After shaping, the assembly was placed in an oven at 70°C for 6 h. The obtained product is composed of 5.3% and 94.7% in volume of foam and wood, respectively. That is to say a volume ratio V_{foam}/V_{wood} of around 0.06.

The morphologies of the final products were determined using a Cambridge Stereoscan S260 scanning electron microscope (SEM). Prior to their analysis, a carbon layer was deposited on the samples. Mapping was realized using the K_α rays of oxygen (0.523 keV), aluminum (1.487 keV), silicon (1.740 keV) and potassium (3.313 keV).

Differential thermal analysis (DTA) and thermogravimetric analysis (TGA) were performed to characterize the thermal properties of the solids. TDA–TGA experiments were carried out in a Pt crucible between 25 and 1200°C using a Setaram Setsys Evolution. The samples were heated at 10°C/min in dry airflow. The sample used for the DTA experiment is taken at the interface between wood and foam.

Mechanical tests in double shear were performed on a Zwick/Roel Z300 load testing machine. The specimens, shown in Figure 1b, were tested with a maximum compression of 50 kN. The measurement of displacement was given by the crosshead displacement. The specimen was placed between two compressive modules, and the inferior modulus was equipped with a ball for better load alignment during testing. The tests in double shear were carried out until sample rupture at a constant displacement velocity of 0.5 mm/min. Behavior curves were produced from the load-displacement records for each material. A correction of the curves at the beginning of each test was performed to compensate for mechanical damages.

RESULTS

To verify the association of the two compounds, durability tests in various solutions were performed (Fig. 2). For the foam, three ranges of pH values were selected. In the presence of a basic medium, the solution reaches equilibrium rapidly at a pH of 11.8, revealing very weak interaction with the foam. When the pH is decreased, equilibrium is reached after 2 min, suggesting that a small amount of basic species are released in the medium, resulting in a final pH value of about 11.8. This value is characteristic of alkaline solutions of carbonate, indicating that some carbonate species are released in the aqueous medium. Moreover, these carbonate species have been observed previously by thermal analysis coupled with MS spectrometry.[5] The presence of carbonate species is more strongly expressed in an acidic medium, where equilibrium between carbonate species and the H^+ from HCl is observed. The basic species tend to govern the overall pH of the medium, as an equilibrium pH of 11.8 is observed even in an initially acidic solvent. The value reaches at high pH value is in agreement with data on geopolymer[6] which corresponds to the natural pH of this compound. At the various pH values, elementary analyses of immersion solutions have been done. The elements extracted from the foam bulk are principally silicon and potassium less than 2%. These experiments indicate that these samples are stable in basic and neutral media whereas they are cleaned in acidic environments due to the carbonate species linked to the potassium present in the samples. It seems that potassium appears a good candidate to promote ion transfer. In the case of wood, the choice of basic solutions such as potassium silicate ($[Si] = 3\times10^{-3}$ mol/L, $[K] = 1.9\times10$ mol/L), sodium hydroxide ($[Na] = 1.02$ mol/L), potassium hydroxide ($[K] = 1.175$ mol/L) and the choice of neutral solutions is realized according to the pH value of the foam mixture, which is around 14. The behaviors (Figure 2b) of solutions in contact with wood were very similar and showed little variation. These experiments show that no damage occurred to foam in contact with a precursor solution.

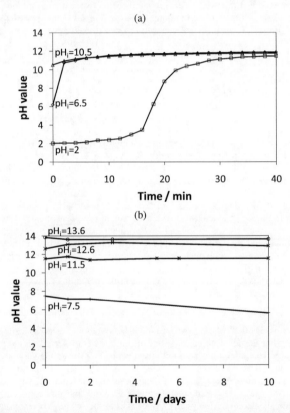

Figure 2: (a) Evolution of solution pH value at various initial pH values in contact with foam and (b) at various initial pH values in contact with wood.

The heat flow curves of wood reveal three separate regions (Figure 3(a)). The first region, from room temperature up to 150°C, corresponds to the dehydratation process.[7] The second region exhibiting exothermic peaks indicates the pyrolysis of cellulose and hemicelluloses, a process which emits CO_2 and CO gas. The exothermic behavior observable in the third region is due to lignin decomposition involving VOC production.[8] The thermal curve of inorganic foam displays two endothermic peaks arising from the loss of dehydrated and structural water. The heat flow curve of the composite appears to be a combination of the two interacting compounds since the maximums of decomposition temperature are shifted to higher temperatures. This effect is in agreement with the work of C. Di Blasi et al., which showed that alkaline compounds highly favour the carbonization, dehydratation and decarboxylation reactions of wood.[9] In these conditions the composite seems to be more stable, demonstrating the existence of a strong interface supporting ion transfer. It is difficult to give a quantitative understanding of the various heat flows. In effect, the weight loss (not given) of the wood is around 88%, around 9% for the foam. The weight loss of the sample tested in DTA allow the evaluation of its composition in terms of wood and foam. At 800°C, wood/foam composite present a weight loss of 24% implying a composition of 20% of wood and 80% of foam. The composite displays only about 20% of the heat flow of the wood.

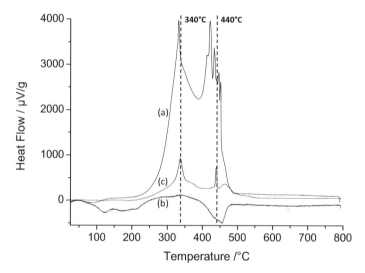

Figure 3: Heat flow as a function of temperature for (a) wood, (b) foam (intensity x10) and (c) wood/foam composites.

Evidence from the durability tests suggests hydric transfer occuring in the interfacial material. In order to identify possible chemical interactions occurring between wood and foam, we performed X-ray mapping at the interface (Fig. 4).

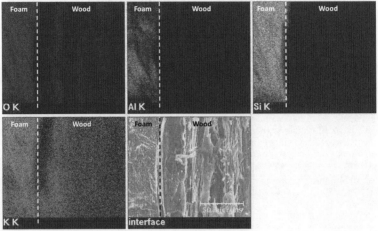

Figure 4: X-ray mapping of oxygen, aluminum, silicon, and potassium realized at the interface between wood and foam.

Additionally, a chemical analysis of wood ash revealed the presence of K, Si, Na, O and C, as expected.[10] The X-ray mapping results revealed a high content of inorganic elements such as Al, Si, O and K in the foam as expected. In the case of wood, the Si, Al and O intensities element are weak since these elements are in weak amount less than 2-3%. However, the strong signal of potassium observed from wood can not be explained by the presence of potassium in wood, but the combination of both potassium in wood and potassium issue from foam. In effect, the inter-diffusion of potassium seems to be responsible of the strong interface in the composite. Indeed, we have previously observed that potassium can migrate easily as potassium hydrogen carbonate by hydration effects which can be promoted at the interface.

(a)

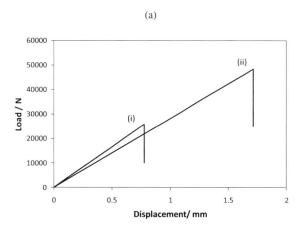

(b)

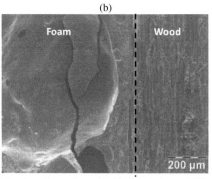

Figure 5. (a) Mechanical behavior of samples with (i) composite and (ii) wood glue layers tested in double-shear. (b) SEM observations of break.

Fig. 5a shows the load-displacement records from the embedding tests. Two kinds of sample have been tested: one with a foam layer and another with an industrial glue layer. The shear failure stress is around 1.3 MPa for the wood/foam composite, and 2.4 MPa for the industrial glue/wood composite. Although the foam/wood composite displays a lower shear failure stress, this mechanical strength is sufficient for composite-containing inorganic materials. In addition, the inorganic foam has interesting adherent properties. SEM measurements (Fig. 5b) provide evidence that mechanical failure due to shear stress occurs in the foam itself and not at the interface. This exemplifies both the brittle behavior of foam and the strong adherence of the foam to the wood substrate.

CONCLUSION

We have demonstrated the development of an insulating geomaterial composite composed of wood and in-situ inorganic foam that can be synthesized at low temperature and with low energy consumption. The interaction between foam and wood leads to the formation of an interface with desirable mechanical properties. These properties combine with the insulating properties of foam which may be of use for constructing fire-resistant doors. Other studies have already shown the possibility of forming inorganic foams with a wide range of clays, especially those which do not require firing. Thus, it is very interesting to investigate the properties of new composites formed with these foams.

REFERENCE

[1] I. Ganguly, C.T. Koebel, and R.A. Cantrell, A categorical modeling approach to analyzing new product adoption and usage in the context of building materials industry, Technological Forescating and social change, 10.1016/j.techfore.2009.10.011 (2009).

[2] F. Shvara, P. Svoboda, J. Dolezal, Z. Bittnar, V. Smilauer, L. Kopecky, and R. Sulc. Geoplymer concrete - An ancient material too?, Ceramics Silikaty, 52 (4), 296-298 (2008).

[3] J.L. Provis, V. Rose, S.A. Bernal, and J.S.J. van Deventer, High resolution nanoprobe X ray fluorescence characterization of heterogeneous calcium and heavy metal distributions in alkali-activated fly ash, Langmuir, 25(19), 11897-11904 (2009).

[4] Buchwald A., Geopolymer binders Part I., ZKG International, 60 (12), 78-84 (2007).

[5] E. Prud'homme, P. Michaud, E. Joussein, C. Peyratout, A. Smith, S. Clacens, J.M. Clacens, S. Rossignol, J. of Eur. Ceram. Soc. 30, 1641-1648 (2010)

[6] Z. Aly, E.R. Vance, D.S. Perera, J.V. Hanna, C.S. Griffith, J. Davis, D; Durce, 2008, Aqueous leachability of metakaolin-based geopolymers with molar ratios of Si/Al=1.5-4, Journal of Nuclear Materials 378, 172-179 (2008)

[7] R. Cerc Korosec, B. Lavric, G. Rep, F. Pohleven, and P. Bukovec, Thermogravimetry as a possible tool for determining modification degree of thermally treated Norway spruce wood, J. Therm. Anal. and Calorim., 98, 189-195 (2009).

[8] M. Jeguirim, and G. Trouvé, Pyrolysis characteristics and kinetics of Arundo donax using thermogravimetric analysis, Bioressources technology, 100, 4026-4031 (2009).

[9] C. Di Blasi, A. Galgano, and C. Branca, Influence of the chemical state of alkaline compounds and the nature of alkali metal on wood pyrolysis, Ind. Eng. Chem. Res.,48, 33559-3369 (2009).

[10] L. Etiegni, A.G. Campbell, Physical and chemical characteristics of wood ash, Bioressource Technology, 37 (2), 173-178 (1991).

EFFECT OF CURING CONDITIONS ON THE POROSITY CHARACTERISTICS OF METAKAOLIN–FLY ASH GEOPOLYMERS

Tammy L. Metroke[1], Michael V. Henley[2], Michael I. Hammons[2]

[1]Universal Technology Corporation, 139 Barnes, Suite 2, Tyndall AFB, FL 32403

[2]Air Force Research Laboratory, Airbase Technologies Division, 139 Barnes, Suite 2, Tyndall AFB, FL 32403

ABSTRACT

The porosity characteristics of metakaolin (MK)- and fly ash (FA)-based geopolymers cured under ambient, thermal, and moist conditions were investigated using nitrogen adsorption porosimetry. In general, surface areas were higher for materials prepared from FA (29.0–59.2 m^2/g) than for MK (5.3–12.8 m^2/g). Total pore volumes ranged from 0.034–0.104 cc/g for MK-based materials and from 0.077–0.089 cc/g for FA-based materials. For MK-based geopolymers, curing under ambient conditions resulted in a single, broad pore size distribution (PSD) between 250–500 Å. PSDs for MK-based geopolymers cured using thermal or moist methods were bimodal with broad peaks between approximately 60–250 Å and 250–500 Å. For FA-based geopolymers, a single, broad PSD was observed and ranged from 20–100 Å for ambient temperature cured materials and 40–200 Å for materials cured under thermal or moist conditions. The observed pore size distributions were accompanied by a shift to higher wavenumber in the position of the main geopolymer band in the ATR–FTIR spectra of geopolymer materials after thermal or moist curing, indicating a higher concentration of Si–O–Si bonds formed in the materials cured at elevated temperatures. These results suggest that the observed PSDs result from Si–O–Al and Si–O–Si bond formation during curing and are affected by the curing conditions.

INTRODUCTION

Curing conditions have been shown to affect mechanical strength development and cracking tendency in geopolymers prepared from metakaolin or fly ash. For example, Palomo *et al.* investigated the effect of curing at elevated temperatures (65–95°C) on the mechanical strength of FA geopolymers[1]. Bakharev showed the beneficial effect of pre-curing at room temperature on the strength properties of FA-derived geopolymers[2]. Kovalchuk *et al.* showed that temperature and humidity play key roles in the development of the microstructure and properties of alkali-activated FA materials[3]. Perera *et al.* investigated the influence of curing schedule on the porosity and cracking tendency of metakaolinite-based geopolymers and found that curing at higher relative humidity does not offer an advantage over curing at ambient followed by heating (40–60°C) in sealed containers[4]. From these studies, it is clear that the effects of curing conditions depend on various factors including source material, temperature, humidity, and curing time and must be considered for each individual geopolymer system.

To complement previous studies, this paper presents a systematic investigation of the effects of curing conditions on the porosity characteristics of geopolymers prepared from MK or FA as determined using nitrogen adsorption porosimetry.

EXPERIMENTAL

Materials

The aluminosilicate source materials were MK (Powerpozz, Advanced Cement Technologies) and class F FA (Boral Material Technologies). MK contained 54–56 % SiO_2 and 40–42 % Al_2O_3. FA contained 53.94 % SiO_2, 28.25 % Al_2O_3, and 7.29 % Fe_2O_3. Sodium silicate D (PQ Corporation) and potassium hydroxide pellets (Aldrich or Fisher) were used. Sodium silicate D contains 29.4 % SiO_2 and 14.7 % Na_2O.

Particle size distributions of the aluminosilicate source materials were determined using an APS 3321 particle size analyzer (TSI Inc.). The mean particle sizes of MK and FA were 0.89 and 1.1 m, respectively. The specific gravities reported by the manufacturer for MK and FA were 2.60 and 2.28 g/cm^3, respectively. The Brunauer–Emmet–Teller (BET) surface areas (0.05–0.3 P/Po) of MK (16.5 m^2/g) and FA (2.3 m^2/g) were measured using an Autosorb-1 porosimeter (Quantachrome Inc.) using N_2 adsorbate at 77 K.

Geopolymer Formulation and Synthesis

Geopolymer compositions were prepared using $SiO_2/Al_2O_3 = 4$ and $M_2O/SiO_2 = 1.75$. The H_2O/M_2O ratio was 11 and 2 for materials prepared using metakaolin and fly ash, respectively. In a typical preparation, sodium silicate and potassium hydroxide were mixed and allowed to cool to room temperature. The solid components were then added with vigorous mixing using a rotary mixer for 10 minutes. The materials were transferred into plastic molds, covered with a plastic top, and allowed to cure at room temperature overnight. Room temperature cured specimens were then allowed to cure, covered, for 28 days. Thermally cured specimens were covered and placed in a 90°C oven for 5 days. Moist cured specimens were placed, uncovered, in an environmental chamber at 52°C and 85% relative humidity for 14 days.

Nitrogen Adsorption

N_2 adsorption/desorption analysis at 77K was performed on powdered specimens using an Autosorb 1 porosimeter (Quantachrome Inc.). Prior to analysis, air/water desorption was typically carried out at 200°C overnight. Surface area was calculated using the BET method in the $0.05 \leq P/P_0 \leq 0.3$ region. Total pore volume was determined at $P/P_0 = 0.995$. Pore size distributions were determined using the adsorption data and the non-local density functional theory (NLDFT) method provided in the ASWin 2.0 porosimeter software. The NLDFT kernel used for data analysis was for N_2 adsorbate on a zeolite adsorbent at 77K.

RESULTS AND DISCUSSION

Nitrogen adsorption/desorption was used to investigate the surface area, total pore volume, and pore size distribution of the MK- or FA-based geopolymer materials as a function of curing method. Figure 1 shows the N_2 adsorption/desorption isotherms of the geopolymers cured under room temperature, thermal, or moist curing conditions. The isotherms were Type IV[5] and revealed capillary condensation accompanied by a hysteresis loop of varying widths, indicating the presence of mesoporosity.

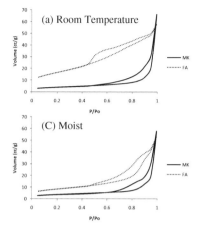

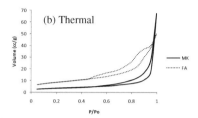

Figure 1: N$_2$ adsorption/desorption isotherms of geopolymers prepared from MK or FA and cured under (a) room temperature, (b) thermal, or (c) moist curing conditions.

BET surface areas were calculated using the linear portion of the isotherm between 0.05–0.3 P/P_0. Figure 2 shows the BET surface area of the MK- and FA-based materials cured under room temperature, thermal, or moist conditions. The BET surface areas were 5.3, 11.9, and 12.8 m^2/g for the MK-based geopolymers cured, respectively, under RT, thermal, or moist conditions. The respective BET surface areas for the FA-based geopolymers cured under room temperature, thermal, or moist conditions were 59.2, 29.8, and 29.0 m^2/g. The significantly higher surface area for the FA-based materials cured under ambient conditions may indicate a lower extent of polycondensation as compared to MK-based materials. For either MK or FA, the BET surface areas for the thermal- and moist-cured materials were similar.

Total pore volume (TPV) is derived from the amount of nitrogen vapor adsorbed at a relative pressure close to unity by assuming that the pores are then filled with condensed adsorbate in the normal liquid state. Due to the lack of a plateau at high P/P_0, TPVs in this study were calculated at P/P_0=0.995. Figure 3 shows the effect of curing condition on the TPV of the MK- and FA-based materials. For room temperature curing, the TPV of the FA-based geopolymer was significantly higher than that of the analogous MK-based material, presumably due to the

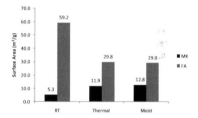

Figure 2: Surface area of MK- or FA-based geopolymers as a function of curing method.

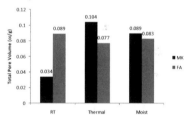

Figure 3: Total pore volume of MK- or FA-based geopolymers as a function of curing method.

more porous, less-condensed nature of these materials. The TPV of materials prepared using thermal or moist curing methods was similar, ranging from 0.077–0.104 cc/g. The TPV for the MK-based geopolymers cured under thermal or moist conditions were higher than for the analogous material cured under ambient conditions, possibly due to a broader distribution of pores.

Non-local density functionally theory is commonly used for the determination of PSD for micro- and mesoporous materials[6]. Figure 4 shows the NLDFT PSDs of the MK- and FA-based materials cured using room-temperature, thermal, or moist curing methods. For the MK-based materials cured under ambient conditions, the PSD was observed in the 250–500 Å region. The PSDs for MK-based materials cured under thermal or moist conditions were bimodal, with broad peaks between approximately 60–250 Å and 250–500 Å. The PSD for FA-based materials cured at room temperature was observed at approximately 20–100 Å. For the FA-based materials, the PSD was observed predominately between 40–200 Å for materials cured using thermal or moist conditions. The observed pore size distributions were accompanied by a shift to higher wavenumber in the position of the main geopolymer band in the ATR–FTIR spectra (not shown) during thermal or moist curing, indicating a higher concentration of Si–O–Si bonds are formed in the materials cured at elevated temperatures.

Geopolymer materials comprise the aluminosilicate gel, unreacted or partially reacted aluminosilicate source materials (amorphous or crystalline), and porosity. The porosity that develops is expected to result from (a) condensation of released $Al(OH)_4^-$ ions with the silicate to form Si–O–Al bonds and (b) silicate–silicate condensation to form Si–O–Si bonds. The three curing methods under investigation involve different combinations of water and thermal contact, influencing the relative rates of dissolution, hydrolysis, and condensation. For MK-based geopolymers cured under ambient conditions, the single PSD between 250–500 Å suggests this peak is due primarily to Si–O–Al bond formation due to silicate–$Al(OH)_4^-$ condensation. The shift in the ATR–FTIR band position of geopolymers cured under thermal or moist conditions indicates that the growth of the broad 60–250 Å peak in the PSD of the MK-based geopolymers is likely due to the formation of Si–O–Si bonds due to thermally-induced silicate–silicate condensation. The single, broad peak in the pore size distribution of FA-based materials cured under thermal or moist conditions may be related to predominately Si–O–Si bond formation through thermally-induced silicate–silicate condensation occurring in the absence of a high rate of release of $Al(OH)_4^-$ ions from FA. To confirm this, additional studies related to the microstructure and connectivity of MK- and FA-based geopolymers as a function of curing conditions will be conducted.

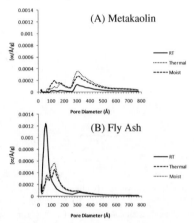

Figure 4: NLDFT pore size distributions for (a) MK- and (b) FA-based geopolymers cured under room temperature, thermal, or moist conditions.

CONCLUSIONS

Porosity characteristics (surface area, total pore volume) and pore size distributions of geopolymeric materials were affected by curing conditions. N_2 adsorption porosimetry showed that MK- or FA-based geopolymers were mesoporous. A shift to higher wavenumber in the position of the main geopolymer band in the ATR–FTIR spectra of geopolymer materials after thermal or moist curing indicated a higher concentration of Si–O–Si bonds formed in the materials cured at elevated temperatures. MK-based geopolymers cured under thermal or moist conditions exhibited bimodal pore size distributions due to silicate oligomer–$Al(OH)_4^-$ condensation (Si–O–Al bond formation) and thermally-induced silicate–silicate condensation (Si–O–Si bond formation). FA-based geopolymers cured under thermal or moist conditions exhibited a single, broad pore size distribution due primarily to Si–O–Si bond formation caused by thermally-induced silicate–silicate condensation and the lower $Al(OH)_4^-$ release rates for FA as compared to MK.

ACKNOWLEDGEMENTS

The authors would like to thank Dr. Ewa Celer for helpful discussions and Ms. Kara Griffith for her assistance with sample preparation.

REFERENCES

(1) A. Palomo, M.W. Grutzeck, M.T. Blanco, Alkali-Activated Fly Ashes: A Cement for the Future, *Cem. Concr. Res.*, **29**, 1323–1329 (1999).
(2) T. Bakharev, Geopolymeric Materials Prepared using Class F Fly Ash and Elevated Temperature Curing, *Cem. Concr. Res.*, **35**, 1224–1232 (2005).
(3) G. Kovalchuk, A. Fernández-Jiménez, A. Palomo, Alkali-Activated Fly Ash: Effect of Thermal Curing Conditions on Mechanical and Microstructural Development-Part II, *Fuel* **86**, 315–322 (2007).
(4) D.S. Perera, O. Uchida, E.R. Vance, Influence of Curing Schedule on the Integrity of Geopolymers, *J. Mater. Sci.*, **42**, 3099–3106 (2007).
(5) K.S.W. Sing, D.H. Everett, R.A.W. Haul, L. Moscou, R.A. Pierotti, J. Rouquérol. T. Siemieniewska, Reporting Physisorption Data for Gas/Solid Systems with Special Reference to the Determination of Surface Area and Porosity, *Pure Appl. Chem.*, **57**, 603–619 (1985).
(6) P.I. Ravikovitch, A. Vishnyakov, A.V. Neimark, Density Functional Theory and Molecular Simulations of Adsorption and Phase Transitions in Nanopores, *Phys. Rev. E.*, **64**, 011601-1–20 (2001).

NEW INSIGHTS ON GEOPOLYMERISATION USING MOLYBDATE, RAMAN, AND INFRARED SPECTROSCOPY.

C. H. Rüscher[1], E. Mielcarek[1], J. Wongpa[2], F. Jirasit[3], W. Lutz[4]

[1] Institute for Mineralogy and ZFM, Leibniz Universität Hannover, Callinstr. 3, 30167 Hannover, Germany
[2] Department of Civil Engineering, Faculty of Engineering, Kings Mongkut's University of Technology Thonburi, Bangmod, Tungkru, Bangkok 10140, Thailand
[3] Institute for building materials, Leibniz Universität Hannover, Nienburgerstr. 3, Hannover, Germany
[4] Brandenburgische Technische Universität Cottbus, Vollmerstr. 13, 12489 Berlin, Germany

ABSTRACT
    Geopolymerisation of an optimally alkali activated metakaolin cement was investigated in dependence on time by strength measurements, by infrared and Raman spectroscopy, and by the molybdate tracer method. The increase in flexural strength at the beginning of aging is explained by the development of two main structural units on different time scales: a fast formation of polymeric silicate chain type (polysiloxo) units and a slow formation of a three dimensional network crosslinking the chains and including sialate bondings. However during further aging a significant weakening occured due to the fragmentation and incorporation of the chain units into the aluminosilicate body. Variations in the waterglass to metakaolin ratio decreasing the nominative K/Al and Si/Al ratio produced about the same binder phase but led to a significant portion of unreacted metakaolin. Further silicate and aluminosilicate cements were synthesized based on rice husk-bark ash, slag, and combinations with slag and metakaolin. It is concluded that the formation and crosslinking of long silicate chains becomes crucial for gaining high mechanical strength and that the protection of the silicate chains becomes crucial for holding long time high strength. This protection is given in the presence of unresolved metakaolin and becomes more pronounced with the addition of significant amounts of CaO together with highly reactive SiO_2 source material (Slag).

INTRODUCTION
    Geopolymers show remarkable properties that make them superior to other working material. One of the most important advantages is there much higher heat tolerance than other organic or inorganic composites. Geopolymer can obtain 70% of the final compressive strength in the first 4 hours of setting. In comparison with concrete made from Portland Cement, geopolymer is characterized by lower environmental impact due to considerably lower production of CO_2 by cement manufacture. One remarkable proof for high durability and in principal the requirement of non-sophisticated preparation techniques could be related to the Egyptian pyramids, which could have been constructed using to a larger extent alkali and earth alkali activated silicate source material as a binder [1]. There are a huge number of investigations concerning geopolymers. One of the most recent reviews has been given in the book edited by J. L. Provis and J. S. J. Deventer [2]. However the basic mechanism of hardening and the main reason for the often reported decrease in mechanical strength with increasing aging time of these silicate and aluminosilicate gels has to our opinion not well been worked out experimentally – at least concerning academic research on metakaolin based geopolymers.
    As written by Breck in 1974 [3] the investigation of aluminosilicate gels could be traced back to J. T. Way, R. Gans, and G. Wiegner in 1850, 1905, and 1914, respectively. Considerable

investigations started during the fifties of the last century in different directions of applications. In one direction the gels were optimised for crystallisation of say microporous and mesoporous materials with first zeolite synthesis using aluminosilicate gels as precursors [4, 5]. Systematic investigations on formation in respect to application as adsorbents, catalysts, ion-exchanger are described by Breck [3], Shdanov [6], and Barrer [7]. As example regarding to the aim of synthesis of zeolite A, X, and Y the chemistry of alumosilicate gel aging was intensively investigated [8-10]. The so called molybdate method was extremely helpful which was originally developed already in 1965 by Thilo et al. [11] for the determination of the degree of polymerisation of silicate anion complexes with working examples of mono-, di- and tetrameta-silicates, phyllosilicates, sodiumsilicate glasses and certain forms of calcium silicate hydrates (CSH). Another advantage in this field was that the structure type, and empirically the Si/Al ratio of the product, e.g. faujasite (FAU) type zeolite, could be obtained using XRD and infrared spectroscopy [12-14]. Moreover the development of high resolution, ^{27}Al and ^{29}Si MAS NMR spectroscopy techniques [15-17] fundamentally improved the knowledge about the relative distribution of the [SiO_4] and [AlO_4] units in the framework, although certain data obtained by XRD, IR and NMR requires some calibration by chemical methods in the presence of superposition effects of framework and non-framework [SiO_2] and [AlO_2^-] species [18].

Important steps for the better understanding of the high strength properties of CSH type phases as obtained using Portland cement can be related to the successful use of IR and NMR techniques [19, 20]. Basically the crystallographic solution of tobermorite crystal structure reveals its characteristic Si-O infinite "Dreiereinfachketten" as embedded within the CaO layers [21, 22], which can be seen to provide high mechanical strength. However, concerning the X-ray amorphous solid aluminosilicate gel it was also realized that typically broadened ^{29}Si and ^{27}Al NMR signal intensities covers all the relevant (SiO_4)-nSi(OAl)$_n$ units [23, 24]. Here one of the main difficulties could be seen for the structural characterisation of geopolymers and their optimisation for binders gaining high mechanical strength, i.e. another direction of applications of alumino-silicate gels. The problem may have been brought to the point by MacKenzie [25] reviewing in 2003 the knowledge about the structure of geopolymers under the title "What are these things called Geopolymers? A Physico Chemical perspective". Briefly geopolymers are synthetic silicate and aluminosilicate gels obtained by geopolymerisation, i.e. by the activation of appropriate source material with alkaline solution and importantly with the addition of waterglass when used as binder. Historically the possibility of preparing binders by the reaction of aluminosilicate raw material (slag, fly ash, clay) with alkaline solutions (hydroxide silicates) goes back to Glukhowski [26]. The term "geopolymer" was introduced by Davidovits [27, 28]. Since then the "geopolymer science" has been promoted at the "Geopolymer Institute" in France and a number of papers appeared and conferences has been held in the field of geopolymers as binders, including the annual meetings within the American Ceramic Society from 2003 on. Nevertheless for an overview of the basic ideas, literature and patents up to 2000 in this field we may refer to the article of Davidovits included in the Geopolymere '99 Proceedings [29]. The nomenclature introduced by Davidovits has been used to describe the hardening mechanism of aluminosilicate gels (geopolymerisation). According to this the polymeric structure results from condensation and cross linking of poly-siloxo chains (-Si-O-Si-O-...) with the so called sialate link – Si-O-Al-O into a three dimensional network. Consequently an empirical formula for polysialate reads $M_n\{-(SiO_2)_z-AlO_2\}_n*wH_2O$, where M is a cation such as potassium, sodium or calcium, which must be introduced for charge balancing of Al^{3+} linked in the network, n describes the degree of polycondensation and z numbers the silicate groups in the chain linked in sialate. The number of water molecules enclosed in cavities of the network, w, can be up to 30 wt%, but unlike their crystalline counterparts, e.g FAU type zeolites, the loss of water during heating occurs irreversibly.

A new aspect for the understanding of geopolymerisation has been reported using the combination of mechanical tests and infrared absorption spectroscopy together with investigations by molybdate measurements [30, 31]. The interpretation of the infrared absorption characteristics during geopolymerisation could still be controversial in regard to earlier reports relating the variation in peak position to the degree of alkali activation [32], to a distinction between alkali activated cements and Portland cements [33] or either to unresolved metakaolin or to the Si/Al ratio [34, 35].

In this work the compressive strength development during aging of two series of mortars and concretes related to metakaolin in combination with CSH type binders and basically high silicate content binder will be considered. It is aimed to better understand the observed behaviour related to a further analysis of data presented in Ref. [30, 31] with particular attention to the acid and alkaline resistivity of aluminosilicate materials. Raman spectroscopy is used as another tool which could provide structural information about geopolymerisation during aging as obtained earlier about variations in the bonding character of silicate groups in solutions depending on molarity [36] or during the synthesis of zeolite Y [37]. The effect on variation in metakaolin to waterglass ratio has been investigated in detail using infrared absorption spectra considering the behaviour in the so called density of states peak maximum (DOSPM) of the asymmetric Si-O infrared active vibrations. The influence of mixtures with slag containing CaO is reported. The DOSPM behaviour during aging of pure silicate based cement using rice husk-bark ash (RHBA) as source material is outlined.

EXPERIMENTAL

Series of cements, Cem.1 - Cem.7, were prepared using a metakaolin Metastar 501 (MK_1) in a solution of potassium water glass Silirit M60 (WG_1). The chemical compositions of MK_1 contains 61.5 % SiO_2, 37.8 % Al_2O_3 and 0.7 % H_2O. WG_1 consist of 19.3 % SiO_2, 24.5 % K_2O and 56.2 % H_2O (all values by weight%). The compositions of the cements were systematically varied with the WG_1/MK_1 ratio and then resulted in the variation of Si/Al, K/Si and K/Al ratios as given in Table 1. The geopolymeric slurry was put into Polyethylene (PE) boxes, closed and aged under ambient conditions for the infrared investigation (KBr method, Bruker IFS66v) and Raman spectroscopy (CRM 200, WITec GmbH Ulm, Germany, using green Laserline). For the molybdate method 5 mg of the sample was suspended in 100 ml HCl (0.01 M) at 273 K and subsequently treated at 298 K with 2 ml of a 10% solution of molybdic acid. The yellow colouration was quantified spectroscopically at 400 nm (Shimadzu). Acid leaching experiments for IR investigations were conducted using 40 mg sample and 0.1 ml 1% hydrochloric acid for 2 h at 4°C. Relative flexural bending force of a Cem.1 composition were measured on 3 cm long and 1.5 cm thick cylindrical pieces on a universal testing machine (MEGA 2-3000-100 D) by a specifically designed device for such sample shapes [30, 31].

Further series of cements were prepared using metakaolin MK_2 (MC-Bauchemie Müller GmbH& Co., Germany) and ground granulated blast furnace slag H (Holcim GmbH, Germany) with compositions as given in Table 2. An alkaline activator were prepared 1 day before mixing using potassium silicate solution with a $K_2O:SiO_2$ ratio of 0.4 (Woellner GmbH and Co. KG, Germany) diluted with an 8 M solution of KOH pellets (Carl Roth GmbH, Germany) obtaining finally waterglass solution with K_2O/SiO_2 ratios $WG_2 = 0.5$, $WG_3 = 0.55$, $WG_4 = 0.6$ (all values by weight). With these waterglass compositions MK_2 based cements Cem.8 - Cem.10 (Table 3) were prepared. Compositions of Cements Cem.11 - Cem.14 including H (ground granulated blast furnace slag) are also given in Table 3. For the IR investigations the slurry were filled into PE boxed, closed and aged at room temperature ambient conditions. For mechanical measurements mortar specimens were prepared with one part of the cements and three parts of washed sand (0.5-2 mm). The mixtures were filled into standard test piece sizes and their mechanical properties were measured in standard testing devices (for details compare Ref. [34, 35]).

Table 1. Chemical composition of Cem.1 – Cem.7 with ratio MK1/WG$_2$ (by weight %) and molar ratio Si/Al, K/Si and K/Al (WG$_2$: weight ratio K$_2$O/SiO$_2$ = 1.27).

Mixture	MK$_1$/WG$_1$	Si/Al	K/Si	K/Al
Cem.2	70/30	2.38	0.72	1.7
Cem.1	62.5/37.5	2.09	0.58	1.2
Cem.3	55/45	1.90	0.47	0.89
Cem.4	48/52	1.78	0.38	0.68
Cem.5	40/60	1.67	0.30	0.50
Cem.6	30/70	1.56	0.18	0.28
Cem.7	25/75	1.52	0.15	0.23

Table 2: Chemical composition of raw materials (by weight %) MK$_2$ and slag (H).

Materials	SiO$_2$	Al$_2$O$_3$	Fe$_2$O$_3$	CaO	SO$_3$	MgO	K$_2$O	Na$_2$O
MK$_2$	49.92	41.45	0.32	0.03	1.76	0.01	0.14	1.32
H	36.52	10.17	0.32	41.61	0.06	5.43	1.06	0.26

Table 3. Molar ratios Si/Al, (K+Na)/Si and (K+Na)/Al of cements 8-10 based on MK$_2$ and WG$_2$ to WG$_4$ and molar ratios Ca/Si, Si/Al, (K+Na)/Si and (K+Na)/Al of H and mixtures of H plus MK$_2$ (cements 11-14). The compositions of the mixtures are given by weight ratios: MK$_2$/H, WG:K$_2$O/SiO$_2$ and WG/Solid. Mortars are related to cements as written in column 2.

Mixture	Related Mortar	MK$_2$/H	WG: K$_2$O/SiO$_2$	WG/ Solid	Ca/Si	Si/Al	(K+Na)/S i	(K+Na)/ Al
Cem.8	Mortar 1	100/0	WG$_2$:0.5	0.75	-	1.55	0.35	0.54
Cem.9		100/0	WG$_3$:0.55	0.75	-	1.51	0.37	0.55
Cem.10	Mortar 1'	100/0	WG$_4$:0.6	0.75	-	1.48	0.38	0.57
Cem.11	Mortar 5	0/100	WG$_2$:0.5	0.75	0.90	4.12	0.31	1.28
Cem.12	Mortar 4	10/90	WG$_2$:0.5	0.75	0.79	3.24	0.30	0.97
Cem.13	Mortar 3	30/70	WG$_2$:0.5	0.75	0.58	2.32	0.28	0.64
Cem.14	Mortar 2	50/50	WG$_2$:0.5	0.75	0.40	1.84	0.26	0.47

CSH type phases were prepared using C$_3$S with the addition of water, which were aged in closed boxes at room ambient conditions for more than 40 days. C$_3$S were synthesized by using commercially available TEOS (C$_8$H$_{20}$O$_4$Si) and Ca(NO$_3$)$_2$•4H$_2$O for gel preparation and final heating at 1400°C.

A pure silicate cement paste has been prepared using rice husk bark ash (RHBA, containing nearly 90 wt% SiO$_2$) from Thai Power Supply (TPS), 14 M NaOH solution prepared from commercial grade NaOH pellets and sodium silicate solution (CR 53, C. Thai Chemicals Co., Ltd.) in weight percent portions 50% RHBA, 35.7 % waterglass solution plus 14.3 % NaOH solution. After mechanical mixing the geopolymeric slurry was filled into PE boxes for the infrared investigation and aged under ambient conditions.

RESULTS AND DISCUSSION

Chemical reactions and structural rearrangements of Cem.1 during aging

Fig. 1 contains the main results concerning flexural force measurements, molybdate measurements, and the density of states peak maximum (DOSPM) of the asymmetric IR active Si-O vibrations of Cem.1. The composition of Cem.1 realizes Si/Al $\approx$ 2 and K/Al $\approx$ 1 in a fully polymerized body with a complete charge compensation of all [AlO$_4$] units by K [30, 31]. This composition was chosen although knowing that a fully polymerized network of Si/Al ratio above about 2.4 could show improved properties e.g. concerning acid resistivity. For example the FAU type structure reveals for Si/Al = 2.4 first infinite long chain type (polysiloxo) silicate units. Smaller Si/Al ratios invariably lead to the breakdown of infinite chain length (i.e. below the percolation threshold) which could be the basis for an improved definition concerning the (hydrothermal) stability ranges of zeolite X and Y [39]. Importantly for the understanding of the results of the molybdate method and also for the understanding of long term mechanical stability is that acid immediately attacks and hydrolysis sialate (Al-O-Si) bondings whereas polysiloxo (Si-O-Si) bondings remain stable or even further polymerise. Contrary to this alkaline solution immediately attacks polysiloxo bondings and stabilizes sialate bondings [40]. With this in mind the 20°C aging behaviour of Cem.1 (Fig. 1) may be considered in detail. A first flexural force could be measured at 24 h (1 d). The values steeply increase to a maximum value after about 96 h (4 d), holding this value up to 192 h (8 d) and then smoothly looses nearly the complete strength towards 528 h (22 d) with a tailing off during further aging (not shown). Similar behaviour was reported earlier by Herr et al. 2004 [41] in relation to the molybdate results.

The initial strong increase in mechanical strength and its crossover into the nearly complete loss must be related to the underlying chemical reactions and structural rearrangements in the gel phase. As shown in Fig. 1 the molybdate active fraction of SiO$_2$ initially decreases to a minimum of about 75 % reached at 24 h of aging, then monotonously increases with a concave-like shape up to about 100 % molybdate activity at 528 h, which implies a most homogeneous fragmentation of silicate units into oligomeric sheet lengths. In other words, the concentration of polymeric silicate units increase up to about 25 % within the first 24 h of aging, and then decrease up to a complete disappearance at 528 h. Polymeric silicate units mean here that they consist of more than 30 units, which are molybdate inactive [11, 18], i.e. the acid attack on these units does not lead to a measurable content of monomeric silicate units. On the other hand a further separation of the molydate active fraction was possible observing a steep increase in molybdate activity which turns over in a more flatter behaviour for all samples investigated. From this a nearly constant contribution of monomeric and dimeric silicate units of about 20 % for aging longer than one day could be deduced. At 24 h of aging a minimum contribution is observed. The decrease in molybdate activity for aging above 528 h indicates a further separation into Si enriched and Al enriched fractions, like a disproportionation of the alumosilicate fragments driven by the requirement that an aluminosilicate gel becomes most stable for a Si/Al=1 ratio.

The infrared spectrum obtained for the sample aged for 1 h reveal a very broad peak in the range of the asymmetrical vibration DOSPM at about 1025 cm^{-1}. The peak maximum further decreases to 1007 cm^{-1} at 13 h and then shifts back to about 1025 cm^{-1} at 24 h. It can be assumed that during this time all the silicate units of the metakaolin becomes largely dissolved which may have completed already at 13 h of aging. This process consumes hydroxide solution taken from the potassium water glass. Therefore, the water glass solution must have condensed, suggesting the formation of mostly chain-like units of polymeric length during further aging up to about 24 h. This effect leads to the shift in the DOSPM up to about 1025 cm^{-1} as observed for aging at 24 h. The increase in concentration of polymeric silicate units is obtained in the molybdate activity, too (Fig. 1). Infrared absorption spectra of condensates from waterglass, obtained by precipitation with drops of HCl, also showed a shift towards higher wavenumbers starting from about 990 cm^{-1} for monomeric silicate unity (Q^0), to 1000 cm^{-1} for dimeric (Q^1) and at about 1010-1020 for chain type Q^2 units. Interestingly it should be noted that Q^4(nAl) units with n = 0, 1, 2, 3, 4 closely coincide with Q^0, Q^1, Q^2, Q^3, Q^4 units in the IR absorption of asymmetric vibration [30, 31].

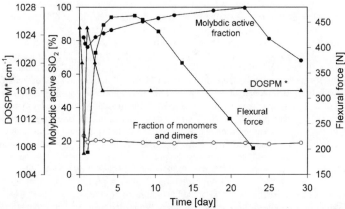

Figure 1. Density of states peak maximum (DOSPM) of asymmetric IR active Si-O vibrations, molybdate active fraction and fraction of monomers and dimers of SiO$_2$ units, and flexural force date of Cem.1 during aging.

Above 24 h a systematic shift in the DOSPM occurs to 1020 cm^{-1} after 48 h, and further to 1016 cm^{-1} after 72 h. Further aging leaves the DOSPM largely invariant (Fig. 1). It can be assumed that for aging more than 24 h a significant amount of silicate units becomes crosslinked including sialate bondings probably involving to a large extent monomeric and dimeric silicate units, their content necessarily remains constant. However, the increasing molybdate activity indicates structural changes which decrease the amount of polycondensed silicate units continuously during aging from 24 h to 528 h (Fig. 1). This indicates that these very long chains become shorter. The density of states of asymmetric Si-O vibrations, although showing constant DOSPM during aging above 100 h, covers changing contents in the degree of condensation of silicate units as obtained by further acid leaching experiments. A minimum shift in the DOSPM for acid leached samples is observed for samples aged around 500 h (Fig. 2). This observation is in line with the results obtained by the molybdate method observing the highest contribution of molybdate activity at 528 h of aging. It may be noted that acid treatment within the molybdate method is followed by complexation of all silicate forming monomers which leads to the characteristic yellow colour. On the contrary, in conducting the acid leaching for the infrared absorption experiment it is the degree of polymerisation of silicate units which lead to the shift in the DOSPM. Any meaningful changes in the DOSPM can only be seen in kinetically controlled leaching experiments interrupted at short times. In the limit, at longer time of acid treatment the DOSPM always tends to 1080 cm^{-1} for all samples irrespective of aging time. Therefore a minimum shift in the DOSPM for the kinetically controlled acid treated samples indicates here the highest stability against acid which can be assumed for a most suitable protection of sialate Al-O-Si bondings within an aluminosilicate network. There are also certain changes in the line shape of the density of states envelope function which could be recovered in second derivative function [38]. However, a meaningful peak deconvolution is presently not possible.

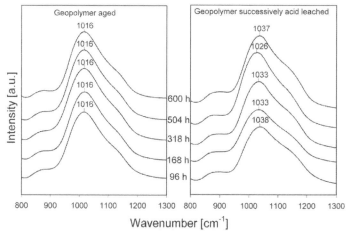

Figure 2. Change in infrared absorption characteristic in the range of DOSPM for acid leached samples in kinetically controlled experiments (right hand) compare to spectra obtained before the leaching (left hand) for Cem.1.

SEM investigations showed for Cem.1 a very homogeneous surface which does not indicate any visible changes depending on aging time. EDX measurement gave an average molar ratio of Si/Al of about 1.9. This value is in good agreement with the nominal molar ratio Si/Al=2.09 obtained from the chemical reactants. This may also indicate the dissolution reaction of almost the whole portion of metakaolin. The DOSPM of 1016 cm^{-1} is also consistent with an average Si/Al ratio of about 2 using an empirical relation DOSPM = 257x+1105 obtained for aluminosilicate glasses [38, 34] where x denotes the Al molar fraction, having in mind that a most homogeneous distribution of Al and Si as indicated for aging of about 528 h.

The variation of intensities during aging of the Si-OH (880 cm^{-1}) and H_2O (1630 cm^{-1}) related peaks were evaluated. The strong effect of network condensation could be observed obtaining a strong decrease in the ratio of Si-OH/H_2O peak intensities during aging up to about 100 h (Fig. 3). According to this on average no further condensation is effective for aging of Cem.1 longer than 100 h which implies that condensation reaction and destruction cancel each other for longer times of aging. It can be assumed that the network condensation sets free KOH solution, where K-ions are not required for charge compensation of sialate bondings. The KOH solution "attacks" polymeric silicate units leading thus to their fragmentation into oligomeric forms. Thus network formation can at some critical degree of fragmentation not any more compensate for the loss in macroscopic strength.

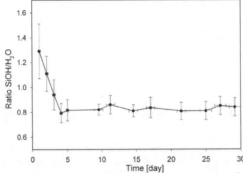

Figure 3. Variation in peak intensity ratio corresponding to Si-OH ($880\ cm^{-1}$) and H_2O ($1630\ cm^{-1}$) during aging of Cem.1.

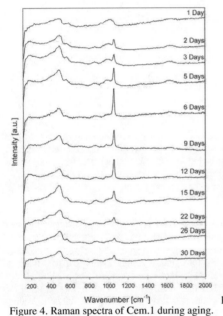

Figure 4. Raman spectra of Cem.1 during aging.

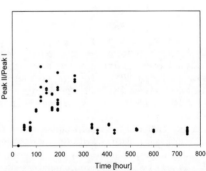

Figure 5. Variation in Raman intensity ratio of peak II/peak I of Cem.1 during aging (taken for peak I at $1064\ cm^{-1}$, peak II at $480\ cm^{-1}$, compare Fig. 4).

    The Raman spectra of Cem.1 (Fig. 4) are dominated by a strong and broad peak with maximum at $480\ cm^{-1}$ (peak II) and a sharper one at about $1064\ cm^{-1}$ (peak I) and some weaker one in between, two of which are most pronounced around $580\ cm^{-1}$ and $890\ cm^{-1}$. For the sample aged 24 h the $1064\ cm^{-1}$ peak is absent and the broader peak with maximum at about $1030\ cm^{-1}$ becomes more visible, which remains as a low wavenumber shoulder to the peak at 1064 cm^{-1} during further aging. The peak at 480 cm^{-1} (peak II) is related to a symmetric motion of bridged oxygen in the Si-O-Si bond [42], whereas the peak at 1064 cm^{-1} (peak I) is related to a Si-O$^-$ vibration where O$^-$ denotes a non-bridging oxygen within a Q^3 or Q^2(1Al) species. A similar peak at 1057 cm^{-1} has also been assigned to a Si-O- related vibration in Q^3 type configuration of aqueous tetramethylammonium silicate solutions by Dutta and Shieh [36]. The true intensity of this peak was difficult to measure. For example the intensity increases with the time of laser illumination, probably related to heating effects. As we measured the same peak position for alkali activated metakaolin using KOH and NaOH related waterglass solutions we rule out its assignment to carbonate species. Additionally plotting the intensity ratio peak I/peak II for all data obtained for type Cem.1 during aging (Fig. 5) revealed a close similarity to the functional dependence of the flexural force (Fig. 1). Thus the increase in intensity of the peak at 1064 cm^{-1} may be related to the crosslinking of silicate chains, revealing a high number of non-bridging oxygens which becomes reduced in content in the course of further rearrangement towards homogenous aluminosilicate fragments.

Changing the waterglass to metakaolin ratio and the degree of alkali activation

For the metakaolin used in the preparation of Cem.1-Cem.7 small amounts of impurity phases could be observed using standard X-ray powder diffraction methods. Such phases obviously doesn't react during geopolymerization, according to the presence in the final products. Metakaolin itself revealed a broad peak centered around 25° (2 Theta). For Cem.1 a broad peak centred around 27-29° (2Theta) is observed. A systematic decrease of the waterglass to metakaolin ratio for Cem.3 to Cem.7 leads to the broadening of the X ray diffraction peak due to a systematic increase in the amount of unreacted metakaolin. The same conclusion can be drawn from the infrared absorption spectra as shown in Fig. 6 for 1 day and 20 day aged samples. A lower waterglass to metakaolin ratio (Tab. 1) leads to systematic higher DOSPM values compared to Cem.1. On the other hand, Cem.2 which consists of a higher waterglass to metakaolin ratio showed a lower DOSPM (1014 cm^{-1}) cured for 1 day and a slightly higher DOSPM (1018 cm^{-1}) compared to Cem.1 indicating a somewhat different behaviour. However, the spectral characteristics with the structure-related peaks DOSPM, around 440 cm^{-1} (Si-O-Si bending), 590 cm^{-1} (indicative for Q^2 type species, as observed in condensed waterglass, but not in fully polymerised silica, [38]) and 700 cm^{-1} (Al-O, Si-O related symmetrical stretching) are clearly observed for Cem.1 to Cem.4 after 1 day of aging as well as after 20 days. For Cem.5 this still holds for the 20 day aged sample as well, whereas for the 1 day sample superimposition becomes more likely implied by the additional shoulder observed at about 780 cm^{-1}. For Cem.3 to Cem.7 the effect of unreacted metakaolin superimposes the binder spectrum the lower the waterglass to metakaolin ratio is and the less the time of aging is. We tried to subtract the metakaolin contribution from the spectra of the 20 day aged sample. Very good agreement to the spectra of Cem.1 was obtained assuming a contribution of 16%, 17%, 35 %, 55 % and 68% for Cem.3, Cem.4, Cem.5, Cem.6 and Cem.7, respectively.

For Cem.1-Cem.7 the KOH content was held constant in the waterglass solution used for alkali activation (Tab. 1). Therefore, the K/Al ratio decreases systematically with decrease in the waterglass to metakaolin ratio. In another series of experiments the KOH content in the activator solution was changed as given in Table 3 (Cem.8-Cem.10). The spectra of the samples with smaller KOH content show only very slight changes in the spectral shape during aging up to 3 days (Fig. 7). A peak

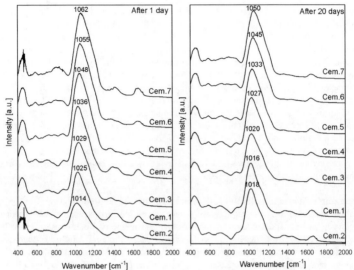

Figure 6. IR absorption spectra of Cem.1 – Cem.7 taken after 1 day (left) and 20 days of aging.

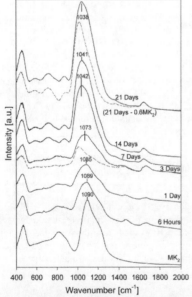

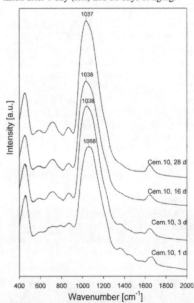

Figure 7. IR absorption of Cem.8 aged as denoted. Dashed curves: x•MK₂ spectrum subtracted.

Figure 8. IR absorption spectra of Cem.10 of aged samples as denoted.

broadening in the range of DOSPM is observed, which shifts to about 1073 cm^{-1}. Subtracting an appropriate contribution of the spectrum of raw metakaolin reveal a significant DOSPM at about 1010 cm^{-1}, a remaining peak at about 440 cm^{-1}, but largely not further resolvable peaks were observed in between. On the contrary the characteristically well defined peaks around 590 cm^{-1}, 700 cm^{-1} and 860 cm^{-1} also seen for Cem.1-Cem.4 are observed for the samples aged above 7 days and longer. The DOSPM has shifted now below 1042 cm^{-1}. Subtracting some appropriate contribution of raw metakolin (e.g. 50-60%) revealed a shift of DOSPM to about 1016 cm^{-1} and a very similar total spectral shape to Cem.1. This could indicate a similar structure of the binder in both cases. For the mixture using a higher KOH content in the activator (Cem.10, Fig. 8) a similar behaviour is observed. However, already the 3 day aged sample showed a strong change in spectral shape and strong decrease in DOSPM compared to the 1 day aged sample

The effect of the different degree of alkali activation on mechanical properties has been checked by setting measurements and compressive and flexural strength measurements to the cement related mortars (Fig. 9, mortars 1, 1'). It is observed that a slightly higher content of KOH within the waterglass leads to a significantly faster setting and faster increase in flexural and compressive strength, gaining however the same absolute values for aging longer than 14 days. Contrary to Cem.1 (Fig. 1) it can be concluded that the relevant Cem.8 – Cem.10 reveal no significant loss in strength. We argue that the portion of unreacted metakaolin mainly works as a buffer, attracting all the KOH which are set free during network formation, contrary to the situation in Cem.1 where KOH leads to the fragmentation of polymeric chains.

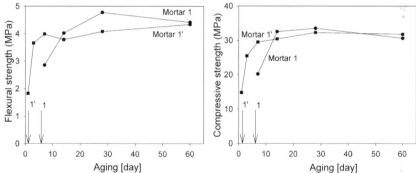

Figure 9. Flexural and compressive strength of Mortars 1 and 1' (Tab. 3) for alkali activations using a higher (Mortar 1') and a lower (Mortar 1) KOH content in the waterglas solution (details are given in the text). The obtained setting times are marked by arrows.

DOSPM behaviour and mechanical properties of alkali activated metakaolin with the addition of CaO-rich slag

For better comparison the results obtained by Jirasit et al. [35] for the DOSPM during aging up to 28 days are replotted here in Fig. 10 for Cem.8 and Cem.11-Cem.14. It can be seen that the influence of slag is to systematically reduce the DOSPM at every time of aging. We argue that the behaviour could be similar to that expected like a mechanical mixing of both reactants and that reaction runs more or less separately as for the constituents alone during alkali activation.

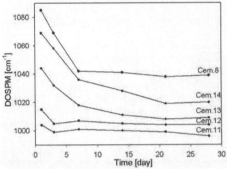

Figure 10. DOSPM behaviour of Cem.8 and Cem.11-Cem.14 (taken from Ref. 35).

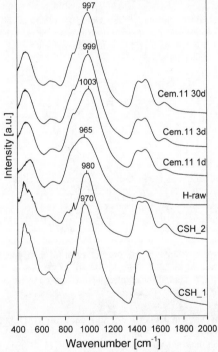

Figure 11. Infrared absorption spectra of CSH type phases of an almost completely reacted C_3S phase with water compared to alkali activated slag samples aged as denoted.

A distinction of geopolymer samples, geopolymer alkali-activated slag samples, alkali-activated slag and ordinary Portland cement samples with peak positions at 1028 cm^{-1}, 1002 cm^{-1}, 988 cm^{-1} and 972 cm^{-1}, respectively, were given related on the specific contribution of Q^n and $Q^4(nAl)$ units [33]. For comparison infrared absorption spectra of CSH-type phases and spectra obtained for alkali activated slag during aging are shown in Fig. 11. According to Yu et al. [19] the 1.1 nm tobermorite posses a sharp peak at about 980 cm^{-1} with shoulders at 1060 cm^{-1} and 900 cm^{-1} which belong to the asymmetric Si-vibrations in mainly Q^2 sites. The peak at 1200 cm^{-1} was assigned to Q^3 sites, since a substantial intensity was observed in ^{29}Si MAS NMR signals related to Q^3 groups. These are absent in 1.4 nm tobermorite were instead Q^1 signals were reported. For 1.4 nm tobermorite the peak shape of the asymmetric Si-O stretching vibration appeared to be broadened and the "DOSPM" appeared to be shifted by about 10 cm^{-1} towards lower wavenumber compared to the DOSPM for 1.1 nm tobermorite. CSH type phases, which were obtained from C_3S material, while the reaction of H_2O shows the DOSPM typically in the range between 980 and 970 cm^{-1} (Fig. 11). It was reported that the hydration products of C_3S reveal CSH phases with C/S ratios of about 1.8 containing only Q^1 and Q^2 units [43]. Thus the spectra shown can be taken as typical for the appearance of a mixture of Q^1 and Q^2 type species in CSH-type phases of high Ca/Si ratio. The small peak at 870 cm^{-1} and the double peak between 1400 and 1500 cm^{-1} is related to carbonation effect under atmospheric conditions. The effect of alkali activation of raw slag now is to produce a DOSPM at about 1003 cm^{-1} which decreases slightly in position during aging. Thus the increasingly lower DOSPM with the increasing amount of slag can be explained by the increasing contribution of CSH type phases obtained by alkali activation.

The compressive strength of Cem.11-Cem.14 corresponding mortars 2-5 with the addition of various amounts of slag are shown in Fig. 12. Mortar 1, based on purely alkali activated metakaolin is shown again. A strong improvement in the compressive strength development is given with the

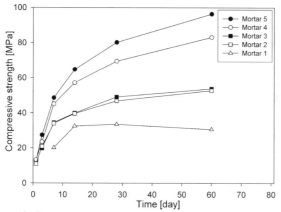

Figure 12. Compressive strength behaviour of mortars 1-5 during aging.

addition of slag containing high amounts of CaO (Table 2). All these mortars show an increase in strength over all times of aging measured in so far (60 days). The absolute values are the higher the higher the slag content, i.e. purely alkali activated slag reaches a value of about 98 MPa at the 60 days testing time. The higher strength of the slag based geopolymers is explained by the formation of CSH type phases obtained by alkali activation. The increase in strength during the whole period of aging (Fig. 12) is related to a further growth of silicate type chains during network formation indicating that Ca could behave as a type of SDA (structure directing agent).

Strength and DOSPM behaviour of alkali activated RHBA (rice husk-bark ash) based binders
Compressive strength behaviour of concrete samples based on RHBA with the addition of fly ash show a similar behaviour also observed for mortar 1 [38] (Fig. 12). Maxima in compressive strength occur after about 14 days with values between about 20 and 34 MPa. The ratio SiO_2/Al_2O_3 of the cement part of these concretes ranges between 5.5 and 6 (Si/Al = 2.75-3) and with one example even up to about Si/Al = 18, where a much smaller amount of fly ash was added. These concretes reach above about 20 GPa after 10 days and continue to increase up to 28 GPa at about 28 days, and then

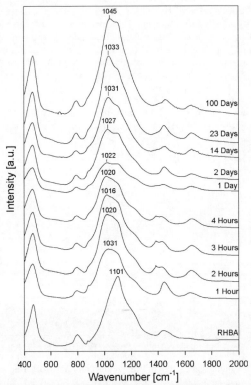

Figure 13. Infrared absorption spectra of alkali activated RHBA during aging as denoted.

very slightly decreases during further aging. Related to this the infrared absorption spectra of a cement prepared by alkali activation of pure RHBA obtained during aging are shown in Fig. 13. During the first 24 h a peak evolves in the DOSPM with a maximum at about 1010-1020 cm^{-1}. During further aging a gradual shift in the DOSPM occurs to about 1045 cm^{-1} after 100 days. The shoulder with a maximum at about 1100 cm^{-1} can be related to unreacted RHBA. Thus the shift of DOSPM from about 1022 cm^{-1} (1 day aging) to higher wavenumbers may be related to the crosslinking of the chain type units (increasing the intensity of Q^3, Q^4 type species). This implies a similar reaction mechanism also seen for alkali activated metakaolin. There occurs a fast condensation of polysiloxo-type units with the extraction of NaOH from the waterglass followed by a slow formation of the network. Here the network does not contain significant contents of sialate bondings, so that the DOSPM shifts towards higher wavenumbers indicated by Q^3 and Q^4 units.

SUMMARY AND CONCLUSION

It has been shown that the key in understanding geopolymerisation is that the waterglass activator looses more or less rapidly its KOH (NaOH) content depending on the solubilty of the silicate or aluminosilicate source. This enforces condensation of the silicate units from the waterglass. Network formation due to crosslinking of silicate chains follows more slowly. As shown for Cem.1 strong structural rearrangements occur during network condensation which leads finally to the breakdown even of the total strength initially gained due to fragmentation of polysiloxo units. Thus protection of the polysiloxo chains is required for long term durability. Such protection could be given by the presence of unreacted metakaolin or with CaO providing tobermorite-type environments for the silicate chains. Spectroscopic evidence for this effect could be given for Cem.11 – Cem.14 (Fig. 14, below).

The results concerning the DOSPM behaviour during aging are summarized in Fig. 14 for direct comparison for the 20 day aged samples. For Cem.1-Cem.7 (black dots) the KOH content was held constant in the waterglass solution used for alkali activation (Table 1). Therefore, the K/Al ratio

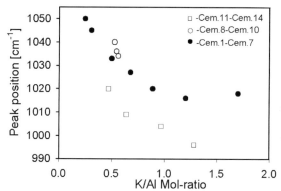

Figure 14. DOSPM of Cem.1-7 (black dots), Cem.8-Cem.10 (open circles) and Cem.11-14 (open squares) related on K/Al ratio. (Compare Tab. 1 and 3, and details given in the text).

decreases systematically with decreasing the waterglass to metakaolin ratio. There is a systematic decrease in values from about 1050 cm^{-1} (Cem.7) to 1016 cm^{-1} (Cem.1) with increasing K/Al ratio which is related to the decreasing content of unreacted metakaolin. For Cem.2 the DOSPM shows a bit higher value at 20 day of aging (1018 cm^{-1}) and somewhat lower value at 1 day of aging (1014 cm^{-1}) which can be related to the higher Si/Al ratio and the higher degree of alkali activation for this composition, respectively. This effect becomes more pronounced for the behavior of alkali activated RHBA (Fig. 13), observing an increase in DOSPM from about 1022 cm^{-1} to 1045 cm^{-1} during aging. Examples obtained in the DOSPM by a systematic decrease in the content of KOH in the activator solution and only a small change in the total Si/Al ratio (Cem.8 – Cem.10) are shown by open circles in Fig. 14. It was observed that only a slight decrease in KOH content led to significantly higher DOSPM during the first days of aging, i.e. the activation of metakaolin occurs rather slowly for Cem.10 (Figs. 8, 9). Finally the systematically lower values of DOSPM for the Cem.11-Cem.14 (open squares in Fig. 14) compared to waterglass-metakaolin based cements can be related to the effect of Ca on the formation of polysiloxo chains.

ACKNOWLEDGEMENT
The authors thank W.M. Kriven (University of Illinois at Urbana-Champaign) for critical reading of the manuscript and giving valuable comments. Parts of the results were contributed from Ph.D related works of EM, FJ and JW. EM thanks for the financial support within "Lichtenbergstipendium" of the "Land Niedersachsen". JW acknowledge support from Thailand research fund (TRF) under the Royal Golden Jubilee Ph.D Program and for the one year stay abroad in the Department of Mineralogy at the Leibniz University of Hannover, Germany. FJ thanks the DAAD for financial support.

REFERENCES
[1] J. L. Provis, J.S.J. van Deventer (ed.), Geopolymers – Structure, processing, properties and industrial applications, Woodhead Publishing Limited, 1-463 (2009).
[2] M. W. Barsoum A. Ganguly, G. Hug, Microstructural Evidence of Reconstituted Limestone Blocks in the Great Pyramids of Egypt, *J. Am. Ceramic Soc.*, **89**, 3788-3796 (2006).
[3] D. W. Breck, *Zeolite Molecular Sieves*, Wiley, New York, (1974).
[4] R.M. Milton, USP 2 882 243, (1953).
[5] R.M. Milton, USP 2 882 244, (1953).
[6] S.P. Zhdanov, S.S. Chvoshchev, N.N. Samulevich, *Synthetic Zeolites* (in Russian), Isd. Chimija, USSR, (1981).
[7] R.M. Barrer, *Hydrothermal Chemistry of Zeolites*, Academic Press, London, (1982).
[8] P. Fahlke, W. Wieker, H. Fürtig, W. Roscher, R. Seidel, Untersuchungen zum Bildungsmechanismus von Molsieben, *Z. f. Anorg. und Allg. Chemie*, **489**, 95-102 (1978).
[9] W. Wieker, B. Fahlke, On the reaction mechanism of the formation of molecular sieves and related compounds, *Stud. Surf. Sci. Catal.*, **24**, 161-181 (1985).
[10] B. Fahlke, P. Starke, V. Seefeld, W. Wieker, K.-P. Wendlandt, On the intermediates in zeolite Y synthesis, *Zeolites*, 209-213 (1987).
[11] E. Thilo, W. Wieker, H. Stade, Über Beziehungen zwischen dem Polymerisationsgrad silicatischer Anionen und ihrem Reaktionsvermögen mit Molybdänsäure. *Z. f. Anorg. und Allg. Chemie*, **340**, 261-276 (1965).
[12] E. M. Flanigen, Zeolite Chemistry and Catalysis, ed. J.A. Rabo, American Chemical Society, Washington, DC, 80-117 (1976).
[13] H. Fichtner-Schmittler, U. Lohse, H. Miessner, H. E. Manek, Correlation between unit-cell parameter, skeletal stretching vibrations and molar fraction of aluminium of faujasite type zeolites for Si:Al = 1.1 – 1000, *Zeitschrift für physikalische Chemie*, Leipzig, **271**, 69-79 (1990).
[14] C. H. Rüscher, J.-Chr. Buhl, W. Lutz, Determination of the Si/Al ratio of faujasite type zeolites, ed. A. Galarneau, F. Di Renzo, F. Fajula, J. Vedrine, Studies in surface science and catalysis, **135**, 1-10 (2001).

[15]G. Engelhardt, U. Lohse, E. Lippmaa, M. Tarmak and M. Magi, ^{29}Si-NMR-Untersuchungen zur Verteilung der Silicium- und Aluminiumatome im Alumosilicatgitter von Zeolithen mit Faujasit-Struktur, *Z. Anorg. Allg. Chem.*, **482**, 49-64 (1981).

[16]G. Engelhardt, U. Lohse, V. Patzelova, M. Mägi, E. Lippmaa, High resolution ^{29}Si n.m.r. of dealuminated Y-zeolites. 1. The dependence of the extent of dealumination on the degree of ammonium exchange and the temperature and water vapour pressure of the thermal treatment, *Zeolites*, **3**, 233-238 (1983).

[17]G. Engelhardt, U. Lohse, V. Patzelova, M. Mägi, E. Lippmaa, High resolution ^{29}Si n.m.r. of dealuminated Y-zeolites. 2. Silicon, aluminium ordering in the tetrahedral zeolite lattice, *Zeolites*, **3**, 238-243 (1983).

[18]W. Lutz, C. H. Rüscher, D. Heidemann, Determination of the framework and non-framework [SiO$_2$] and [AlO$_2$] species of steamed and leached faujasite type zeolites: calibration of IR, NMR and XRD data by chemical methods. *Microporous Mesoporous Mater.*, **55**, 193-202 (2002).

[19]P. Yu, R. J. Kirkpatrick, B. Poe, P. F. McMillan, X. Cong, Structure of calcium silicate hydrate (C-S-H): near-, mid-, and far-infrared spectroscopy. *J. Am. Ceram. Soc.*, **82**, 742-748 (1999).

[20]P. Colombet, A.-R. Grimmer (ed.), Nuclear Magnetic Resonance Spectroscopy of Cement based Materials, Springer Verlag Berlin, (1997).

[21]S.R. Hamid, The crystal structure of the 11 A natural tobermorite Ca$_{2.25}$(Si$_3$O$_{7.5}$(OH)$_{1.5}$)•1H$_2$O, *Z. Kristallographie*, **154**, 189-198 (1981).

[22]S. Merlino, E. Bonaccorsi, T. Armbruster, Thze real structure of tobermorite 11 A: normal and anormalous forms, OD character and polytypic modifications, *Eur. J. Mineral*, **13**, 577-590 (2001).

[23]G. Engelhardt, B. Fahlke, M. Mägi, E. Lippmaa, High resolution solid state ^{29}Si and ^{27}Al n.m.r. of aluminosilicate intermediates of zeolite A synthesis. Part II, *Zeolites*, **3**, 292-294 (1983).

[24]G. Engelhardt, B. Fahlke, M. Mägi, E. Lippmaa, High resolution solid state ^{29}Si and ^{27}Al n.m.r. of aluminosilicate intermediates in the synthesis of zeolite A, *Zeolites*, **5**, 49-52 (1985).

[25]K.J.D. MacKenzie, What are these things called geopolymer? A Physico-Chemical perspective, *Ceramic Transaction-Advances in Ceramic Matrix Composites IX*, **153**, 175-186 (2003).

[26]Glukhovsky, V. D., A new building material, *Bulletin of Technical Information*, Glavkievgorstroy, N2, (1957).

[27]J. Davidovits, Solid phase synthesis of mineral blockcomposite by low temperature polycondensation of alumino-silicate polymers. *IUAPC Macromolecular Symp.*, Stockholm, Sweden, 3p (1976).

[28]J. Davidovits, Synthesis of new high-temperature, Geo-Polymers for reinforced plastics/composites. *SPE PACTEC'79, Society of plastic engineers*, Brookfield Center, USA, 151-154 (1979).

[29]J. Davidovits, Chemistry of Geopolymeric Systems", Terminology. In: J. Davidovits, R. Davidovits, and C. James editors. GÉOPOLYMÈRE '99 International Conference, France, 9-40 (1999).

[30]C.H. Rüscher, E. Mielcarek, W. Lutz, A. Ritzmann, W.M. Kriven, The aging process of alkali activated metakaolin. Ceramic Transactions, **215**, in press.

[31]C.H. Rüscher, E. Mielcarek, W. Lutz, A. Ritzmann, W. M. Kriven, Weakening of alkali activated metakaolin during ageing investigated by the molybdate method and infrared absorption spectroscopy, J. Am. Ceram. Soc., 1-6 (2010), in press, online available.

[32]W.K.W. Lee, and J.S.J. van Deventer, Use of infrared spectroscopy ro study geopolymerisation of heterogeneous amorphous aluminosilicates, *Langmuir*, **19**, 8726-8734 (2003).

[33]I. Lecomte, C. Henrist, M. Liegeois, F. Maseri, A. Rulmont and R. Cloote, (Micro)-structural comparison between geopolymers, alkali-activated slag cement and Portland cement, *J. Europ. Ceramic Soc.*, **26**, 3789-3797 (2006).

[34]F. Jirasit, C.H. Rüscher, L. Lohaus, A study on the substantial improvement of fly ash based geopolymeric cement with the addition of metakaolin. International Conference on Pozzolan, Concrete and Geopolymer Khon Kaen, Thailand, 24-25 May, 1-15 (2006).

[35]F. Jirasit, C.H. Rüscher, L. Lohaus, Property development of the slag and fly ash based geopolymeric cement. International Conference on Sustainability in the Cement and Concrete Industry, Lillehammer, Norway, Norwegian Concrete Association, 149-162 (2007).

[36]P.K. Dutta, D.C. Shieh, Raman spectroscopic studies of aqueous tetramethylammonium silicate solutions, *J. Raman Spectrosc.*, **16**, 312-314 (1985).

[37]P.K. Dutta, D.C. Shieh, M. Puri, Raman spectroscopic study of the synthesis of zeolite Y, *J. Phys. Chem.*, **91**, 2332-2336, (1987).

[38]C.H. Rüscher, E. Mielcarek, J. Wongpa, C. Jaturapitakkul, W. Jirasit, L. Lohaus, Chemical and mechanical properties of silicate and aluminosilicate gels for building materials. European Journal Mineralogy. in preparation.

[39]C.H. Rüscher, N. Salman, J-Chr. Buhl, W. Lutz, Relation between growth-size and chemical composition of X and Y type zeolites. Letter to the editor, *Microp. Mesop. Mat.*, **92**, 309-319 (2006).

[40]N. Salman, C. H. Rüscher, J-Chr. Buhl, W. Lutz, H. Toufar, M. Stöcker, Effect of temperature and time in the hydrothermal treatment of HY zeolite, *Microp. Mesop. Mat.*, **90**, 339-346 (2006).

[41]R. Herr, W. Lutz, A. Ritzmann, B. Hillemeier, K. Schubert, Tagung Bauchemie Erlangen, Gdch. Monographie Bd., **31**, 78-88 (2004).

[42]P. McMillan, B. Piriou, A. Navrotsky, A Raman spectroscopic study of glasses along the joins silica-calcium aluminate, silica-sodium aluminate, and silica-potassium aluminate. *Geochimica et Cosmochimica Acta*, **46**, 2021-2037 (1982).

[43]W. Wiecker, C. Hübert, D. Heidemann, Recent results of solid state NMR investigations and their possibilities of use in cement industry. Int. Congress on Cement Chemistry. Gothenburg (Sweden) ISBN91-630-5495-7, (1997).

TRANSFORMATION OF POLYSIALATE MATRIXES FROM AL-RICH AND SI-RICH
METAKAOLINS: POLYCONDENSATION AND PHYSICO-CHEMICAL PROPERTIES

Elie Kamseu, Cristina Leonelli
Department of Materials and Environmental Engineering, University of Modena and Reggio Emilia,
Via Vignolese 905/A, 41100 Modena, Italy
Mail: kamseuelie2001@yahoo.fr

ABSTRACT

Two metakaolins from Al-rich and Si-rich kaolinitic clays were used to design polysialate matrices
with different Si/Al and NaK/Al ratios. The aim was to investigate the influence of oligomers formed
during dissolution and hydrolysis on the polycondensation and transformation to hard and stable
matrices. Products of geopolymerization of the different matrices were subjected to mechanical testing
considering various loading configurations. The geopolymer matrices showed compressive strength
from 51 ± 5 MPa (Si/Al = 1.23) to 61 ± 2 MPa (Si/Al = 2.42) and bi-axial four-point strength from 11
± 2 MPa to 16 ± 1.1 MPa respectively. These results obtained were consistent with density, leaching
ability, and microstructure. It is proposed that the mechanical properties and the stability of the
products of reactions can be discussed as the important parameters for the evaluation of geopolymer
matrices. Moreover, polycondensation and the final performance of the product are greatly influenced
by the unreacted crystalline or semi-crystalline phases that act as fillers and contribute to increase the
stability and mechanical properties.

1. INTRODUCTION

Geopolymerization occurs at much higher solid/liquid ratios, with high viscous paste with incomplete
dissolution of solid, where precipitation and polymerization of dissolved species are highly significant
in the reaction process[1-10]. From the mixing of alkaline solution with metakaolin, there is a period of
time necessary to form stable nuclei so that network can begin. This period of time for metakaolin vary
from 2 to 6 hours depending on the Si/Al ratio, the degree of amorphization, water, impurities content,
etc… The solution activator will also play important role. Increasing the Si/Al molar ratio generally
decreases the initial rate of the reaction, with the highest Si/Al samples showing what appears to be a
pause in the reaction corresponding roughly to the solidification of geopolymer binder[10]. The above
parameters make the transformation of metakaolin to geopolymer complex with the necessity to master
each components with its influence on the system as well as each sequence of the synthesis that will
accordingly influence the final product.

During the transformation of kaolin to metakaolin, the hydroxyl groups are removed from kaolinite octahedral layers in a layer-by-layer sequence, thus satisfying the first-order kinetics postulated by Brindley and Nakahira[11]. The nature and grade of kaolin as well as the temperature of calcination will affect this transformation. The transformation will influence the silicate and alumina oligomers formed during the geopolymerization. The resultant Al and Si species from metakaolin in alkaline conditions are $[Al(OH)_4]^-$, $[SiO(OH)_3]^-$ and $[SiO_2(OH)_2]^{2-}$ with the concentration ratio of $[SiO_2(OH)_2]^{2-}$ to $[SiO(OH)_3]^-$ increasing with pH value. The silicate species developed will vary from $[Si_4O_8(OH)_6]^{2-}$, at 6-10 M, to $[SiO(OH)_3]^-$ at 11-13 M of NaOH or KOH solutions[1,3].

Independently of the Si/Al molar ratio, the microstructure of the matrices formed will show inhomogeneity since the quick reaction between $[Al(OH)_4]^-$ and $[SiO(OH)_3]^-$ is followed by the slow binding of $[Al(OH)_4]^-$ to the product of reactivity, when Si/Al $\leq$ 1, or by slow reactions solely involving silicates species, when Si/Al > 1. The aim of this work is to understand how these reactions in the case of Si-rich or Al-rich metakaolin affect the properties of geopolymers that we will measure through the porosity, microstructure, densification and mechanical properties. In the development of dense matrices with various physico-chemical phenomena already described[1,3,6,9] the reaction of dissolution and that of polycondensation cannot be separated. The gel formation and polycondensation reactions are simultaneous. The assessment of the extend of dissolution, gel formation and polycondensation influence the final properties of the products obtained. Moreover, the chemical stability and the management of micro cracks can be used to define a good geopolymer. The aim of this work is to understand how these reactions in the case of Si-rich and Al-rich metakaolin affect the properties of geopolymers that we will evaluate through the density, microstructure, water absorption and mechanical properties.

2. EXPERIMENTAL
Two kaolinitic clay materials were chosen based on their alumina and silica content: one with Si/Al0 1.15 called Al-rich, and the second with Si/Al > 2.19 called Si-rich (Table 1). The two materials were from Mayouom (MAY) and Ntamuka (TAN) in the west region of Cameroon[12]. They were dried, calcined at 700°C for 4 hours and ground to fine powders (ϕ < 80 µm).

Five compositions of amorphous alumino-silicates were prepared from the two metakaolins to have different Si/Al molar ratios of 1.15, 1.40, 1.59, 1.86 and 2.19. The molar ratios 1.15 and 2.19 represent 100 wt% of MK-MAY and MK-TAN respectively. The molar ratios 1.40, 1.59 and 1.86 correspond to the partial replacement of MAY with 25, 50 and 75 wt% TAN (Table 1).

Table 1: Physical Properties and Mineralogy of two kaolinitic clays used.

	Mineral phases	L.O.I (wt%)	Particles >100μm	Particles < 40μ m	Particles < 10 μm	Specifique Area m²/g	Si/Al
MAY	kaolinite, Illite, ilmenite, quartz	13.90	0	84.97	56.29	17.10	1.15
TAN	kaolinite, quartz	11.80	0	92.08	65.52	24.32	2.19

Discs of 25 mm (diameter) x 60 mm (thickness) were prepared for compressive strength testing. Others with 40 mm of diameter and 7 mm thickness for bi-axial four-point flexural bending strength. Samples of 140 mm x 10 mm x 10 mm were prepared for uni-axial four-point flexural strength testing. The samples were directly sealed from atmosphere. After 72 hours following the moulding, samples were cured at ambient temperature for at least 10 days before characterization.

To control the stability of the products, after exposition to atmosphere following complete curing, specimen of each composition was cut to $2.0 \pm 0.1g$. The specimens were introduced separately into 100 ml of distillated water in a Becker. The pH of water was analyzed after 10 min, 24hours and 48 hours. The specimen was washed separately with distillated water and being reintroduced again in new distillated water (the same volume). The pH is analyzed again after 10 min, 24 and 48h.

The compressive strength was determined by using testing machine type MTS 810, USA. The end surfaces of specimens were polished flat and parallel to avoid the requirement for capping. The discs were centred in the compression-testing machine and loaded to complete failure. The compressive strength was calculated by dividing the maximum load (N) at failure by the average cross-sectional area (m²). All the values presented in the current work were an average of five samples, with error reported as standard deviation from the mean. The same machine with changing in loading configuration was used for the uni-axial four-point flexural bending strength testing. The sample holder has a span between the two bearers of 42 mm (L2). The distance between the two loading pistons was 18 mm (L1). Supports and both loading pistons were steel knife edges, rounded to a radius 3.05 mm.

Mineralogical analysis of the metakaolins and geopolymer specimens were carried out with an X-ray powder diffractometer (XRD), CuKα, Ni-filtered radiation (the wavelength was 1.54184Å), Phillips Model PW3710. The radiation was generated at 40 mA and 40 kV. The analysis was performed in fine grains of ground metakaolin. For the geopolymers, pieces of materials obtained after compressive strength testing were ground to fine powder. Specimens were step-scanned as random powder mounts from 5 to 70° 2theta at 0.05 2theta steps and integrated at the rate of 2s per step.

The microstructure of the geopolymer specimens was studied using an Environmental Scanning Electron Microscope (ESEM), Model Quanta 200. Samples were coated in thick layer of resin and polished using consecutively finer media up to finer diamond paste of 1μm. To ensure the conduction of electron during the microstructural analysis and the image stability, specimens were gold-coated with 10 nm thickness. The ESEM was coupled with an Oxford Instruments energy dispersive spectrometer (EDS) for the microanalysis that permit the investigation on phases distribution in the matrixes.

3. RESULTS AND DISCUSSION

Figure 1 present the variation of pH of different compositions of geopolymer materials in aqueous medium. From the results, the pH varies with time and composition.

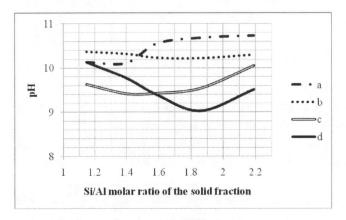

Figure 1: Variation of the PH of geopolymer materials in the aqueous medium a) 10 min after immersion, b) 24 h after, c) samples have been removed from the water after 24h washed before reimmersion for 10 min, d) reimmersion for 24h and 48h.

The samples GP-M and GP-75M did not show important variation after washing but they maintained high values of pH. The Si/Al molar ratios in the final products were 1.23 and 1.55 with 0.83 and 0.89 NaK/Al molar ratios. After washing, the sample GP-M remain with the pH >10 suggesting that the release in this sample is low but continuous and more than 48h could be necessary for confirmation. For the compositions with high degree of polycondensation and stable phases, these crystals do not affect the final properties and can be easily leached into aqueous solution (GP-MT, GP-25M and GP-T). The geopolymer defined as an inorganic polymer make of undissolved aluminosilicates particles embedded in a newly formed and mainly amorphous gel of significant Al substitution within the silicate network[9-10, 13-15] is in mainly the matrix after a washing following the production. The instability of the final matrix from GP-M, compared to the good properties of others compositions suggesting that Si-rich metakaolin are the most suitable aluminosilicate for the development of structural geopolymer materials while Al-rich metakaolin can be used only with the solution of reinforcement with appropriate fillers or a chemical stabilizer of the excess of $[AlO_4]^-$ oligomers.

The variations of compressive, uni-axial and bi-axial four-point flexural strength are shown in Figure 2. The compressive strength varies from 51 ± 5 MPa for Al-rich metakaolin based polysialates matrix (GP-M) to 61 ± 2 MPa for Si-rich polysialates matrix (GP-T). The intermediate compositions have values in the range: 50 ± 4 MPa for GP-MT and 55 ± 1 MPa for GP-75M and GP-25M. These values are closed to those described in the literature[7-9] and far from those obtained by some authors[5,6]. The compressive strength, between 51 ± 5 and 61 ± 1 MPa agree with values obtained by Duxson et al.[8] and can be compared to that from the studies of Rowles and O'Connor[15]. Our results demonstrated that in the range of Si/Al molar ratios from 1.4 to 2.4, the compressive strength values do not vary extensively. In this interval the number of micro flaws distributed in the matrix should be very low due to the degree of reactivity and the equilibrium that results from the interlocked species.

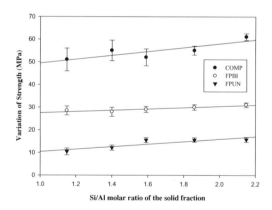

Figure 2: Variation of Compressive (COMP), Bi-axial four-point (FPBI) and Uni-axial four-point (FPUN) flexural strength of geopolymer samples as function of Si/Al molar ratio.

In this range the polysialates formed contribute to coarse the microstructure and various residual particles present as unreacted silicate act as cracks bridging and cracks deflection elements. The matrix present good resistance in aqueous medium, there are low nucleation and growth of micro cracks. These enhance the elastic response of the products and justify the behaviour of those materials in tensile and compressive solicitations(Figure 2).

Figure 3 (a and b) show matrixes with long sheets embedded in a homogeneous amorphous texture for GP-M and GP-75M respectively. The polymers from the two samples could form matrixes primarly of polysialates for GP-M and polysialates with small proportion of polysialate disiloxo for GP-75M. The two matrixes did not have coarse morphology as consequence of the nature of phases present and the absence of more complex polysialates and remaining Si-particles that do not participate to the polysialates formation: GP-M remains with finest texture of the matrix where there can be observed smoothness of the surface. The microstructure remains consistent with the mineralogy of the starting metakaolin. The XRD patterns of MK-M (Figure 4) and GP-M remain primarily amorphous which is in agreement with the theory of geopolymerization defended by many authors: during geopolymerization the dissolution, hydroxylis and nucleation of gel are generally exothermic, afterward the energy of growth of gel to polysialates is not enough to conduct to crystalline phases.

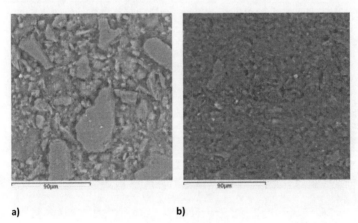

a)   b)

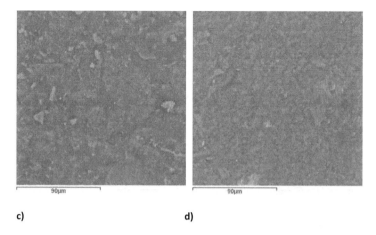

c)   d)

Figure 3: Micrographs of geopolymer samples showing gradation in morphology with Si/Al molar ratio: a) GP-M (Si/Al = 1.23; NaK/Al = 0.83), b) GP-75M (Si/Al = 1.55; NaK/Al = 0.89), c) GP-25M (Si/Al = 2.07; NaK/Al = 1.05) and d) GP-T (Si/Al = 2.42; NaK/Al = 1.16).

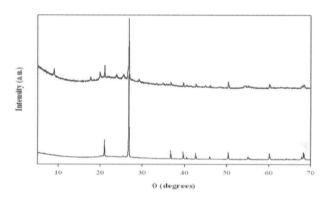

Figure 4: XRD patterns of MK-M and MK-T

The exothermic reactions are at the origin of most of the open pores with large volume observed in the GP-M and GP-75M. Pores that should explain the value of water absorption o figure 5 nonetheless the matrix were found to be relatively compact. The two matrices can be described as the accumulation of small particles that are similar to precipitates with self aggregation.

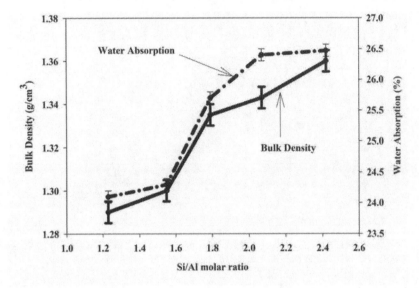

Figure 5: Variation of Bulk Density (g/cm³) and Water Absorption (%) of geopolymer samples as function of Si/Al molar ratio.

By increase the Si-rich in the compositions (GP-MT, GP-25M, GP-T), the variation in the texture was evident with the gradation in the structure types from GP-75M (Figure 3). The matrixes are structured as combinations of amorphous to semi-amorphous blocks compatible with each other but presenting a minor change in composition at local level. It can be suggested that during polycondensation of oligomers formed, any pair of Al and Si oligomers act as nucleation site that will progressively grow by fixation of new particles or cross linking of grains through reinforcement with incongruent dissolved and unreacted silica particles (Figure 3c and 3d, Figure). The presence of extra silica particles in the system enhance the polycondensation of gel in ring and complex structures essentially made on siloxonate and polysiloxonate with low or without Alumina content that will act to reinforce the structure of polysialates already formed and will be at the origin of local inhomogeneity aforementioned. This inhomogeneity also grow with the Si/Al. The particles of Si that do not react and those with incongruent dissolution will remain in the matrix reacting at the surface to consolidate the Si-rich compostions (GP-MT, GP-25M and GP-T) thereby forming part of the structure.

CONCLUSION

In this study, stability after curing in atmosphere and in aqueous medium was combined to the mechanical strength as well as density to investigate on the suitability of Al-rich and Si-rich metakaolin materials for the development of geopolymer matrixes for structural applications. The investigation produced a number of noteworthy conclusions:

Al-rich metakaolin (MK-MAY) to leach in the alkaline solution more $[AlO_4]^-$ than $OSi(OH)_3$ which contribute to extra $[AlO_4]^-$ not comfortably bounded or segregated in the matrix; the final product result instable in atmosphere with time and in aqueous medium with the decrease of mechanical properties (GP-M) as consequence of cracks and fractures development.

Si-rich would leach more $OSi(OH)_3$ than $[AlO_4]^-$ in the alkaline solution which contribute to the extra $OSi(OH)_3$ that will develop in the matrix further polycondensation forming new phases essentially Si-rich that will crosslink in the system through formation of siloxane and sialate links. Good mechanical properties (Compression 60 MPa) is obtained with the material relatively porous but stable at air and in aqueous medium (GP-T). The material present relatively high bulk density compared to Al-rich based polysialate matrixes.

By mixing, in various proportion, both materials, compositions with Si/Al molar ratio in the range of the most stable matrixes (1.4 to 2) were developed (GP-75M, GP-MT and GP-25M) with mechanical properties and density improved compared to standard metakaolin: the compressive strength that was > 50 Mpa alone with the stability of the products in aqueous medium permit to described these compositions as suitable for the structural matrixes.

The choice of Si-rich metakaolin, instead of addition of amorphous silica with different concentration of silicate solution, is then alternative contribution for the strength enhancement.

Both solution and solid-state reactions such as incongruent dissolution of semi-crystalline and crystalline particles must be considered to fully described the polysialate matrixes and their formation.

The mixing of the Al-rich and Si-rich metakaolin also permit to design geopolymer with particular XRD and microstructural features which are characteristic to the mode of polycondensation that take place in those compositions.

REFERENCES

1. L. Weng, K. Sagoe-Crentsil, Dissolution processes, hydrolysis and condensation reactions during geopolymer synthesis: Part I-low Si/Al ratio systems, J. Mater. Sci, 42 (2007)2997-3006.

2. E Kamseu, C Leonelli, D N Perera, U C Melo, P N Lemougna, Investigation of Volcanic ash based geopolymers as Potential Building Materials Interceram 58: 2-3(2009)136-140.

3. L. Weng, K. Sagoe-Crentsil, Dissolution processes, hydrolysis and condensation reactions during geopolymer synthesis: Part II-High Si/Al ratio systems, J. Mater. Sci, 42 (2007)3007-3014.

4. Catherine A. rees, John L. Provis, Grant C. Lukey, Jannie S. J. Van Deventer, The mechanism of geopolymer gel formation investigated through seeded nucleation, Colloids and Surfaces A: Physicochem. Eng. Aspects 318 (2008)97-105.

5. Zhang Yunsheng, Sun Wei, Li Zongjin, Preparation and microstructure of K-PSDS geopolymeric binder, Colloids and Surfaces A: Physicochem. Eng. Aspects 302 (2008)473-482.

6. Hua Xu, J. S. J. Van Deventer, The geopolymerization of alumino-silicate minerals, Int. J. Miner. Process. 59 (2000) 247-266.

7. Hongling Wang, Haihong Li, Fengyuan Yan, Synthesis and mechanical properties of metakaolinite-based geopolymer, Colloids and Surfaces A: Physicochem. Eng. Aspects 268 (2005)1-6.

8. P. Duxson, S. W. Maillicot, G. C. Lukey, W. M Kriven, J.S.J. Van Deventer, The effect of alkali and Si/Al ratio on the development of mechanical properties of metakaolin-based geopolymers, Colloids and Surfaces A: Physicochem. Eng. Aspects 292 (2007)8-20.

9. J. Davidovits, Geopolymer Chemistry and Applications, Publ. Morrisville, USA, 570P; 2008.

10. John L. Provis, Jannie S.J. Van Deventer, Geopolymerisation Kinetics. 1. In situ energy-dispersive X-ray diffractometry, Chemical Engineering Science 62 (2007)2309-2317.

11. P.R. Suitch, Mechanism for the dehydroxylation of Kaolinite, Dickite, and Nacrite from room temperature to 455°C, J. Am. Ceram. Soc., 69 (1986)61-65.

12. E. Kamseu, C. Leonelli, D.N. Boccaccini , P. Veronesi , P. Miselli , Giancar Pellacani , U. Chinje Melo, Characterisation of porcelain compositions using two china clays from Cameroon, Ceramics International 33 (2007) 851–857.

13. D. K. Shetty, A.R. Rosenfield, P. M. Guire, P. Bansal, J. K. Winston, H. Duckeworth, Biaxial flexure tests for ceramics, Amer. Ceram. Soc. Bull. **59** 1193-1197 (1980).

14. F. Zibouche, H. Kerdjoudj, J.B.E de Lacaillerie, H.V Damme, Geopolymers from Algerien metakaolin. Influence of secondary minerals. Applied Clay Science 43 (2009) 453–458.

15. Matthew. R. Rowles, and Brian H. O'Connor, Chemical and Structural Microanalysis of Aluminosilicate Geopolymers Synthesized by sodium Silicate Activation of Metakaolinite, J. Am. Ceram. Soc. 92, 10(2009)2354-2361.

EFFECT OF HIGH TENSILE STRENGTH POLYPROPYLENE CHOPPED FIBER REINFORCEMENTS ON THE MECHANICAL PROPERTIES OF SODIUM BASED GEOPOLYMER COMPOSITES

Daniel R. Lowry and Waltraud M. Kriven
Department of Materials Science and Engineering, University of Illinois at Urbana-Champaign, Urbana, IL USA

ABSTRACT
The geopolymer matrix is ideal for incorporating a variety of reinforcements for structural applications. In the case of this experiment, the reinforcements are high tensile strength polypropylene (PP) chopped fibers of lengths of 0.5", 1", and 2". The fibers were added in increments of 0.92, 1.01, 1.32, and 2.48 weight percents. These samples were then tested in a three point bending apparatus at a loading rate of 176 Newtons per minute. The maximum stress for each sample was then compared to that of the pure, nonreinforced, sodium geopolymer matrix. The samples prepared using the 0.5" fibers achieved average maximum stresses of 9.9±0.5, 9.0±1.3, 9.9±3.0, and 15.1±1.5 MPa, respectively for increasing weight percent of fibers. Samples prepared using the 1" fibers had average maximum stresses of 9.6±3.8, 10.5±1.6, 12.4±1.34, and 14.4±1.0 MPa with respect to increasing fiber contents. The samples prepared with the 2" fibers had average maximum stresses of 9.0±0.9, 11.2±1.4, 13.5±1.7, and 18.3±0.8 MPa with respective increasing fiber composition. This data was compared to the average maximum stress achieved by the pure sodium geopolymer (NaGP) of 1.8±0.5 MPa. From this data Weibull statistics were calculated to determine the mechanical reliability of the material. According to the Weibull statistics for each fiber length used, the 0.92, 1.01, and 2.48 weight percent samples produced the most reliable data, where the 1.32 weight percent samples produced data that was reasonably reliable, while the pure geopolymer samples did not produce reliable data.

INTRODUCTION

Geopolymers are a class of ceramic material first reported by Davidovits in 1978[1]. This material class consists of ceramics derived from a colloidal reaction of aluminosilicates with an activator composed of aqueous hydroxides and silicates. The reaction involves a silicate solution made from sodium hydroxide (NaOH), amorphous silica (SiO_2) and water (H_2O). This is then combined with an aluminosilicate source, such as metakaolin ($Al_2O_3 \cdot 2SiO_2$), the dehydrated form of the kaolinite clay. The geopolymer reaction is as follows:

$$2SiO_2 \cdot Al_2O_3 + Na_2O \cdot 2SiO_2 \cdot xH_2O \rightarrow Al_2O_3 \cdot Na_2O \cdot 4SiO_2 \cdot xH_2O$$

Once prepared the solution can be cast into a desired shape and cured at near ambient conditions in a relatively short time.

The properties of the solid and liquid geopolymer components directly influence the final product. The physical properties of the solid aluminosilicate (e.g. particles size and specific surface area) determine if it will partially or completely dissociate, while the silicate solution will determine if the solid will be partially or completely dissolved. The properties of the silicate solution determine the breakup and recombination of the aluminosilicate structure, as well as the polycondensation and charge balance of the reaction system.

The use of geopolymers for structural applications is limited by the characteristics of the pure geopolymer system. Like other ceramics, geopolymers have their strengths in compression and hardness applications, and they are weak in tensile and flexure loading. To compensate for the inherent weakness in flexure loading, reinforcements can be added, in this case fiber reinforcements. Fiber reinforcements help to improve the flexure properties of the geopolymer sample through crack deflection and crack bridging mechanisms. This study does not address the issue of tensile weakness or any tensile strengthening mechanisms.

EXPERIMENTAL PROCEDURE

Sample Preparation

      Geopolymer samples were made from a sodium metasilicate silicate solution consisting of sodium hydroxide, fumed silica, (Cabot, Tuscola, IL) and distilled water in the proportions shown in Table I, corresponding to molar ratios of $Na_2O \cdot 2SiO_2 \cdot 11H_2O$. The sodium metasilicate solution was combined with metakaolin to synthesize the geopolymer product. By using high tensile strength polypropylene (PP) fibers (Innegrity Inc., Greer, SC) of different lengths, 0.5", 1", and 2", as well as of different amounts of the fibers in the geopolymer matrix, viz., 0.92, 1.01, 1.32, 2.48 wt%, this study could demonstrate the effect of fibrous reinforcement on the mechanical properties of the geopolymer matrix. Figures 1(a) to (c) are SEM micrographs of the fibers, where the images were taken with a JEOL 6060LV SEM. Table II lists the relative proportions of sodium metasilicate solution, metakaolin, and PP fibers constituting each of the samples.

      Samples were prepared by combining the appropriate quantity of each component of the geopolymer mix. First, the metakaolin and sodium metasilicate were mixed to a homogeneous consistency with the use of a high shear mixer (Model RW20DZM, IKA, Germany). Once the geopolymer paste was prepared, the fibers were incorporated by low shear, mechanical mixing. After the fibers had been incorporated the bars were cast in 1" x 1" x 6" plastic molds. The bars were cured in a controlled atmosphere of 50° Celsius and 20 percent relative humidity using a TestEquity model 1007H Temperature-Humidity Chamber (TestEquity, Moorpark, CA). After a period of 24 hours, the samples were removed from the chamber, and demolded, at which point they were ready for testing of their flexure strength in three point bending.

Table I: Composition of Sodium Silicate Solution following 11/1/4 ratio

	Fumed Silica	NaOH	Deionized Water
Amount (grams)	741.63	493.87	1333.42

Table II: Sample Composition of Silicate Solution, Metakaolin, and PP Fibers

Sample	Sodium Silicate	Metakaolinite	PVA Fiber	Wt% PVA
Sample 1	107.09g	58.52g	1.54g	0.92%
Sample 2	107.09g	58.54g	1.69g	1.01%
Sample 3	107.09g	58.55g	2.21g	1.32%
Sample 4	107.11g	58.57g	4.22g	2.48%

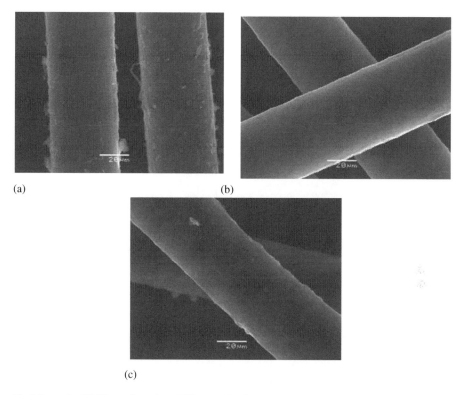

(a)   (b)

(c)

Fig.1 Innegrity PP fibers viewed at x900 magnification (a) 0.5" fibers with diameter 47±1 μm (b) 1" fibers with diameter 50±1.5 μm (c) 2" fibers with diameter 48±1 μm.

2.2 Sample Testing

The samples were tested using ASTM C78 protocols as a guide. The apparatus used for testing was an Instron Universal Testing Machine, Model 5882 with a 50 kN load cell. Samples were broken in a three point flexure configuration with the bottom supports 40 mm to either side of the point of loading on the top surface.

For testing, the samples were turned on their side in relation to their molded position. The samples were not preloaded as prescribed by the ASTM. Since an estimate for the ultimate loading was not known the samples could not be preloaded. The loading rate on the sample was determined by equation 1 as follows:

$$r = \frac{Sbd^2}{L} \tag{1}$$

In this equation, r is the loading rate in Newtons per minute; S is the rate of increase in maximum stress on the tension face in MPa/minute; which was taken to be 0.86 MPa/min; b is the average width of the sample; d is the average depth, where both were taken to be 25.4 mm; and L is the span length which was taken to be 80 mm. From this equation, the calculated loading rate was 176 N/min.

2.3 Sample Analysis
From the observed test data, the modulus of rupture (MOR) in units of Pascals was calculated by equation 2:

$$R = \frac{3PL}{2bd^2}$$ [2]

In this equation, R is the modulus of rupture (in Pa), P is the maximum applied load (in N), L is the span length (measured to be 80 mm). In the denominator, b is the average specimen width at fracture, and d is the average specimen depth at the fracture, both are constant values of 25.4 mm.

After calculating the MOR, the data was plotted against the specimen strain to produce a stress versus strain diagram, which was used to analyze the mechanical properties of the reinforced materials. The MOR is reported as the flexure strength of the samples as these terms are interchangeable.

To produce the Weibull statistics required a series of calculations and data plots, where an approximation of the probability of failure was first determined using equation 3:

$$F \sim \frac{n}{N+1}$$ [3]

where F is the probability of failure, n is the ranking of the sample in its respective sample pool and, N is the number of samples.

From this data, a plot of the natural log of $1/[1-F]$ $\left(\ln\frac{1}{1-F}\right)$ versus the natural log of the

stress ($\ln\sigma$) was plotted and a line of best fit applied to the graph. The slope of this line, m, was then used to approximate the probability of fracture for that sample set. Once determined, this probability was used in equation 4 to determine the effective volume under stress.

$$V_{eff} = \frac{V}{2(m+1)^2}$$ [4]

where v is the volume of the sample (98.3 cm^3), and m is the Weibull Modulus.

After determining the effective volume, the Weibull function can be calculated using equation 5:

$$F = 1 - \exp\left[-V_{eff}\left(\frac{\sigma}{\sigma_o}\right)^m\right]$$ [5]

where F is the Weibull Function, V_{eff} is the effective volume from equation 4, and σ is the applied stress, σ_o is the normalizing stress (when the probability of failure is 0.632) yielding m as the Weibull Modulus.

Determining the standard deviation of the data is important as this value is used to verify the consistency of the data. The data is more replicable when the standard deviation is a low value. The standard deviation was calculated using equation 6:

$$S = [\sum_{i=1}^{N} \frac{(\sigma_i - \bar{\sigma})^2}{N}]^{\frac{1}{2}}$$

[6]

where S is the standard deviation, σ_i is the strength of the individual test bar, $\bar{\sigma}$ is the average strength of the sample set, and N is the number of samples

RESULTS AND DISCUSSION

The three point bending flexure strength of the reinforced geopolymer matrix was determined using an Instron Universal Testing Machine. The resulting data was then quantitatively analyzed. From the measurements of fracture under 3-point loading, the data that was most important to this study was the maximum flexure strength. Comparing this to the natural flexure stress of the geopolymer enables a qualitative comparison of the increase in three point flexure strength of the matrix.

Figure 2 compares the average maximum flexure strength for each class of sample prepared and tested. This includes the three different fiber lengths tested and the different compositions of fibers based on weight percent of the sample. The maximum strength at 0.92 wt% occurred in the 0.5" long fibers while at every other composition the maximum strength was achieved using the 2" long fibers. Maximum three point flexure strengths of 1.76 MPa, 9.96 MPa, 11.22 MPa, 13.63 MPa, and 18.33 MPa were achieved, with respect to the increasing composition of the samples, starting with pure sodium geopolymer.

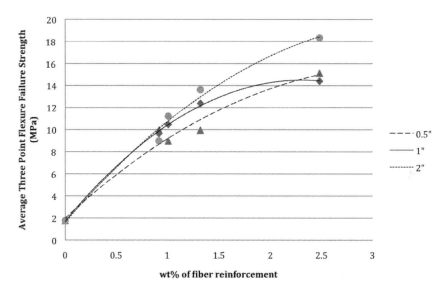

Fig.2 Average maximum failure stress of sample as a function of fiber length and composition.

Figures 3-5 are Weibull function plots of each sample composition using the different fiber lengths. The data for the 0.92, 1.01, and 2.48 wt% composition samples were the most consistent and reliable based on the Weibull function plots. For the 1.32 wt% composition the 1" fiber data is reliable, while the 0'5" and 2" fiber data should be considered but not deemed reliable. Using the Weibull plot of the pure geopolymer samples, the conclusion was made that this material is inherently brittle and the failure point is unpredictable.

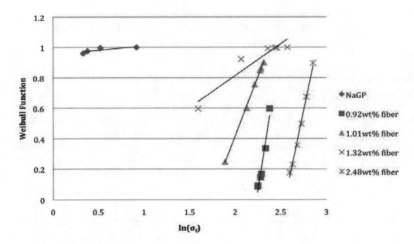

Fig.3 Weibull function *vs* natural log of mean failure stress for 0.5" fibers

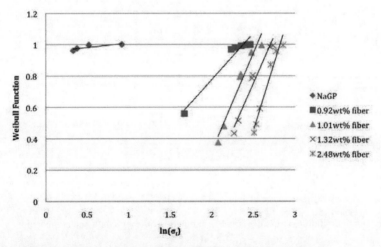

Fig.4 Weibull function *vs* natural log of mean failure stress for 1" fibers

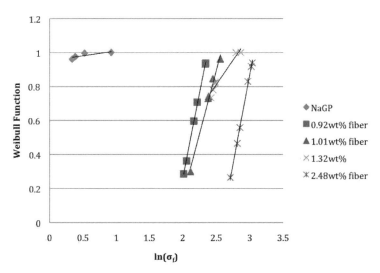

Fig. 5 Weibull function *vs* natural log of mean failure stress for 2" fibers

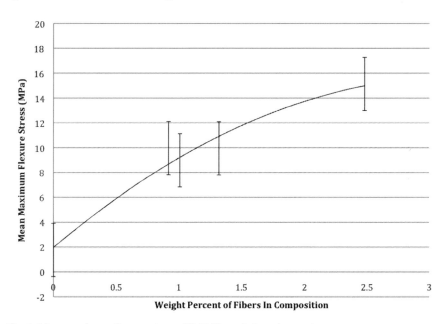

Fig. 6. Mean maximum flexure stress of 0.5" fiber reinforced geopolymer

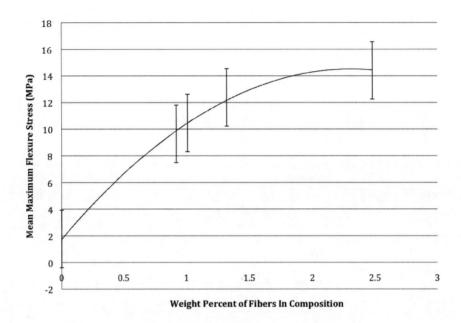

Fig.7. Mean maximum flexure stress of 1" fiber reinforced geopolymer

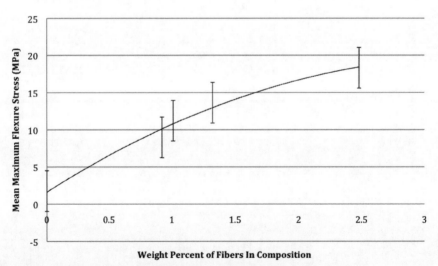

Fig.8 Mean maximum flexure stress of 2" fiber reinforced geopolymer

Comparing the maximum flexure strengths of the reinforced samples to the strength of the pure geopolymer sample showed increases of 466%, 538%, 674%, and 941% for the 0.92 wt% (0.5" fibers), 1.01 wt% (2" fibers), 1.32 wt% (2" fibers), and 2.48 wt% (2" fibers) compositions, respectively. The minimum flexure strength of the reinforced geopolymer compared to the strength of the pure geopolymer indicated percent increases of 410% (2" fibers), 410% (0.5" fibers), 465% (0.5" fibers), and 719% (1" fibers) as a function of increasing fiber composition.

Table II: Statistics for Fiber Reinforced Sodium Geopolymer

Fiber Length	0.5"				
Composition (wt%)	Minimum Stress(MPa)	Maximum Stress(MPa)	Mean Stress(MPa)	Standard Deviation	Effective Volume (cm^3)
0	1.398	2.512	1.765	0.513	3.228
0.92	9.460	10.806	9.959	0.518	0.159
1.01	6.617	10.141	8.980	1.309	1.410
1.32	4.921	13.114	9.944	3.007	4.848
2.48	13.430	17.370	15.126	1.471	0.444

Fiber Length	1.0"				
Composition (wt%)	Minimum Stress(MPa)	Maximum Stress(MPa)	Mean Stress(MPa)	Standard Deviation	Effective Volume (cm^3)
0	1.398	2.512	1.765	0.513	3.228
0.92	5.312	11.732	9.647	2.298	3.895
1.01	7.972	13.471	10.461	2.047	1.580
1.32	9.672	15.433	12.379	2.366	1.537
2.48	12.288	17.390	14.408	2.053	0.895

Fiber Length	2.0"				
Composition (wt%)	Minimum Stress(MPa)	Maximum Stress(MPa)	Mean Stress(MPa)	Standard Deviation	Effective Volume (cm^3)
0	1.398	2.512	1.765	0.513	3.228
0.92	7.472	10.395	8.984	1.230	0.899
1.01	8.308	12.934	11.220	1.715	1.354
1.32	11.258	17.594	13.627	2.686	1.534
2.48	15.040	20.845	18.325	2.313	0.817

Figures 6-8 provide the maximum flexure strengths for each sample composition using the PP fibers as reinforcements, where the error bars indicate the 14th to 86th percentile of data collected during testing. For the samples prepared with 0.5" fibers the increases in flexure strength with respect to increasing fiber content were 464%, 409%, 463%, and 757%. For the samples prepared with the 1" fibers, the increases in flexure strength as a function of increasing fiber content were 447%, 493%, 601%, and 716%. %. For the samples prepared with the 2" fibers the increases in flexure strength as a function of increasing fiber content were 409%, 536%, 672%, and 938% as a function of increasing fiber compositions of the samples. This data and other statistical data for the samples is compiled in Table II above.

CONCLUSIONS

In this study, the fiber reinforced geopolymer samples were prepared by adding PP chopped fibers according to a pre-determined weight percent of the total sample weight. The conclusion was reached that even the smallest addition of reinforcement, dramatically increased the flexure strength of the geopolymer composite. To validate the study of how fibers affect flexure strength controlled quantities of fibers based on weight percent of the total sample were added. A second variable, fiber length, was introduced to investigate its effect on the flexure strength of the geopolymer composite.

The predominant issue with preparing the samples by mechanically adding the fibers to the liquid geopolymer paste was that as more fibers are added, entanglement of the fibers occurs, which complicates the process of producing homogeneous samples. This problem was complicated when using longer fibers, as entanglement became an issue at a smaller addition of the fibers. However, this problem was not significant enough to diminish the results of this study. The Weibull plot for the samples prepared with 2.48 wt% of the 2" fibers demonstrated that the data was reliable for the sample pool tested.

ACKNOWLEDGEMENTS

This work was supported by a Grant from the United States Air Force Office of Scientific Research under Nanoinitiative Grant No. FA 9550-06-1-0221, through Dr. Joan Fuller. In addition, portions of this work were carried out in part in the Frederick Seitz Materials Research Laboratory Central Facilities, University of Illinois, which are partially supported by the U.S. Department of Energy under grants DE-FG02-07ER46453 and DE-FG02-07ER46471. The authors thank Dr. Martin Pfahler of Keanetech,LLC for suggesting the use of the polyproplyene reinforcing fibers and arranging for samples to be graciously supplied by Innegrity Inc., Greer, SC.

REFERENCES

[1]Joseph Davidovits, Journal of Thermal Analysis **1991**, *37*, 1633.
[2]Peter Duxson, John L. Provis, Grant C. Lukey, Seth W. Mallicoat, Waltraud M. Kriven, Jannie S. J. van Deventer, Colloids and Surfaces A: Physicochemical and Engineering Aspects, Volume **269**, Issues 1-3, 1 November 2005.
[3]Hongling Wang, Haihong Li, Fengyuan Yan, Colloids and Surfaces A: Physicochemical and Engineering Aspects, Volume **268**, Issues 1-3, 31 October 2005.
[4]Zhang Yunsheng, Sun Wei, Li Zongjin, Zhou Xiangming, Chau Chungkong, Construction and Building Materials, Volume **22**, Issue 3, March 2008.
[5]Richerson, David W., Modern Ceramic Engineering Second Addition, Marcel Dekker, Inc.

PROPERTIES OF BASALT FIBER REINFORCED GEOPOLYMER COMPOSITES

E. Rill, D. R. Lowry and W. M. Kriven

Department of Material Science and Engineering, University of Illinois at Urbana-Champaign, Urbana, IL, 61801, USA

ABSTRACT

The properties, microstructure, and processing of potassium-based geopolymer ($K_2O \cdot Al_2O_3 \cdot 4SiO_2 \cdot 11H_2O$) composites made with basalt chopped fibers have been systematically studied. Geopolymers were manufactured at ambient temperature and cured in a humidity controlled, constant temperature oven at 50 °C. The effects of varying weight percents of fibers, the presence of a polymer sizing, and processing methods were examined by comparing the strength and work of fracture of each sample with pure geopolymer. Basalt chopped fibers with a length of ¼ inch increased the strength of a bend bar broken in a three point flexural test from an average of 1.7 MPa to an average of 19.5 MPa. Weibull statistics were analyzed and indicated a systematic and significant improvement in reliability due to the addition of chopped basalt fibers.

INTRODUCTION

Geopolymers are a class of amorphous aluminosilicate materials, composed of cross-linked alumina (AlO_4^-) and silica (SiO_4) tetrahedra to form polysialates, with an alkali metal ion to balance the negative charge. In this experiment, potassium is the balancing cation. The term "geopolymer" was first coined in 1976 by Davidovits to describe the "soil cements" of the time. Geopolymers are formed by reacting a metakaolin ($2SiO_2 \cdot Al_2O_3$) with an alkaline solution of the balancing cation of choice. Previous papers by Kriven et al.[1-21] have shown cause to believe that a composition with 11 moles of water per mole of 1.2 m sized metakaolin allows the majority of the materials to react and polymerize. As such, the composition for this paper was therefore fixed at $K_2O \cdot Al_2O_3 \cdot 4SiO_2 \cdot 11H_2O$.

As the geopolymer matrix is heated up to about 350- 400 °C the water in the geopolymer structure escapes from the bulk of the material which is accompanied by mechanical deformation and cracking. Since this cracking can often be catastrophic, and while geopolymers otherwise have good high-temperature qualities, it is of high value to research methods of reducing or eliminating this cracking. One of these ways is with fiber reinforcement, such as with basalt fibers.

The basalt fibers were provided by Basaltex Inc., and the tows were 0.25 inches long with individual fiber diameters of 13 μm. Basalt is a naturally occurring material, composed (at. %) primarily of silica and alumina with a nominal composition of 57.5 % SiO_2, 16.9 % Al_2O_3, 9.5 % Fe_2O_3, 7.8 % CaO, 3.7 % MgO, 2.5 % Na_2O, 1.1 % TiO_2, and 0.8 % K_2O. It has a melting point of 1450 ± 150 °C and is non-combustible, making it useful for high temperature applications. The silane coating or "sizing" helps to protect the brittle fibers from premature fracture, and prevents them from binding to each other.

EXPERIMENTAL PROCEDURES

In order to make the geopolymer matrix, a potassium hydroxide solution (~5 molar solution of KOH) was first mixed with metakaolin clay using an IKA shear mixer (Model RW20DZM, IKA, Germany) to ensure a low viscosity, homogenous slurry. The final geopolymer had a composition of $K_2O \cdot Al_2O_3 \cdot 4SiO_2 \cdot 11H_2O$. In this experiment the amount of water was kept constant ($H_2O/K_2O = 11$), as previous papers by Kriven et al.[1-2] suggested that this concentration provides the least amount of unreacted phase left in the geopolymer after curing. The basalt fibers were then mixed into the liquid geopolymer by hand. The shear mixer proved to destroy the brittle fibers, chopping them into lengths on the order of 100 microns. The Thinky® centrifugal mixer also caused the same effect, leaving mixing to be done manually in order to ensure that the fibers did not fracture.

After the fibers were thoroughly distributed within the geopolymer matrix, the mixture was poured into custom Teflon molds, 6 inches in length with a 1 inch square cross-section. The molds were sealed using Saran® wrap to maintain the correct amount of water in the geopolymer (by preventing water loss during curing). The samples were cured at 50 °C in a constant-humidity oven for 24 hours. The mold was then disassembled in order to prevent pre-cracking in the samples during removal.

All samples were subjected to a three point flexural test, administered using an Instron Universal Testing Frame, following ASTM standard C 78-09. The heat treatments, if applied, were done in a small, calcining furnace using a Eurotherm 2400 series temperature controller. The samples were heated to the target temperature (500 °C or 1000 °C) at a ramp rate of ±5 °C/min, with a soak time at the target temperature for 1 hour.

RESULTS AND DISCUSSION

Amount of Reinforcement Phase

ASTM standard C 78-09 was modified for the specific geometry of the samples, viz., a 6 inch long bend bar with a 1 inch square cross section. The two supports for the sample were placed 80 mm apart, with the load applied equidistant between them. Figures 1-6 show the flexural stress versus flexural strain for each of the amounts of fiber reinforcements tested (excluding the heated samples), including pure geopolymer (Fig. 1). The pure geopolymer had a low average flexural strength of 1.7 MPa (5 samples) (Fig. 1), while the strongest set of samples, 10 % reinforcement (by weight) had an average of 19.5 MPa (6 samples) (Figs 2 to 6). 10 wt % began to reach the approximate limit to the amount of reinforcement phase due to issues with workability and creating a homogeneous mixture.

1 %, 3 %, 5 %, and 7 % of chopped fiber additions (all by weight) showed improved strengths over the pure geopolymer as well (Figs. 1-6). All six sets of samples showed some variability in strengths, though nothing too extreme. However, certain sets contained only 5 samples, as the 6th sample either differed significantly from the geometry required by the test or had significant cracks present from processing (usually upon removal from the molds). That being said, geopolymers are a ceramic-based material, and as such, flaws play an extremely important role in determining how the samples fracture. The samples had relatively low variability which is thought to be due to the fibers bearing much of the load, but the averages stated previously did not encompass the occasional outlier.

To produce the Weibull statistics required a series of calculations and data plots, where an approximation of the probability of failure was first determined using the equation:

$$F \sim \frac{n}{N+1}$$

where F is the probability of failure, n is the ranking of the sample in its respective sample pool and, N is the number of samples.

From this data, a plot of the natural log of $1/[1-F]$ divided by ($\ln 1/(1-F)$) versus the natural log of the stress ($\ln \sigma$) was plotted and a line of best fit applied to the graph. The slope of this line, m, was then used to approximate the probability of fracture for that sample set. Once determined, this probability was used in the equation to determine the effective volume under stress.

$$V_{eff} = \frac{v}{2(m+1)^2}$$

where v is the volume of the sample (98.3 cm^3), and m is the Weibull Modulus.

After determining the effective volume, the Weibull function can be calculated using the equation :

$$F = 1 - \exp\left[-V_{eff}\left(\frac{\sigma}{\sigma_o}\right)^m\right]$$

where F is the Weibull Function, V_{eff} is the effective volume from equation 4, and σ is the applied stress, σ_o is the normalizing stress (when the probability of failure is 0.632) yielding m as the Weibull Modulus.

Determining the standard deviation of the data is important as this value is used to verify the consistency of the data. The data is more replicable when the standard deviation is a low value. The standard deviation was calculated using the equation:

$$S = \left[\sum_{i=1}^{N} \frac{(\sigma_i - \bar{\sigma})^2}{N}\right]^{\frac{1}{2}}$$

where S is the standard deviation, σ_i is the strength of the individual test bar, sigma bar is the average strength of the sample set, and N is the number of samples. The results of the Weibull analysis are shown in Figs. 7. The average 3-point flexure strength as a function of basalt fiber loading is shown in Fig. 8.

In general, the higher percentages of fibers yielded the higher flexural strengths, as well as allowed them to withstand larger strains. While all the fractures seen were catastrophic, the addition of even 1 wt % of fibers did provide a small, though noticeable, additional work of fracture. The peak stress-strain curves for the chopped fiber-reinforced samples were much more rounded than the relatively sharp peak of the pure geopolymer samples. The chopped fiber-reinforced samples also could withstand small amounts of load after full fracture as the fibers nearest the point of load application had not been pulled out of the matrix. This fiber pullout also added to the work of fracture, and its effect increased with the amount of. Overall, the samples with the larger amounts of fiber reinforcement showed higher strengths and work of fracture, as expected. By increasing the amount of reinforcements, the geopolymer matrix was required to carry less and less of the applied load, and crack length increased dramatically, as did the toughness due to fiber pullout.

Basalt Fibers without Sizing

The basalt fibers from Basaltex Inc., come coated with a proprietary silane coating, which helped to maintain the fiber strength while improving workability by preventing them from adhering to each other. Removing the coating (by calcining the fibers at 600 °C for 1 hour) reduced the average strength from 3.6 MPa to 2.4 MPa for a 1 wt % reinforcement phase (Fig. 9). This suggested that the coating did in fact help to maintain fiber strength and should be left on the basalt fibers for maximum composite strength.

Dehydration

As a preliminary investigation into the beneficial effect of basalt chopped fibers on preventing dehydration cracking in the geopolymer matrix, two samples with 7 wt % fiber reinforcement were heated for 1 hour, one at 500 °C and the other at 1000 °C. After heating and upon visual inspection, two changes from the pre-fired state were readily apparent: microcracking was present in the sample and was visible without the aid of magnification, and the samples also appeared to have shrunk. The microcracking was expected, as the dehydration is known to cause cracking as the water molecules attempt to force their way out of the geopolymer structure, though less destructive overall than samples tested in other experiments. The volume change was more noticeable in the 1000 °C sample, and is thought to be due to the crystallization of the matrix from the amorphous geopolymer phase to the glass-ceramic leucite phase, as the transformation occurred just above 1000 °C. However, XRD and SEM analysis will be required to confirm the presence of leucite, in the basalt-containing samples.

The 500 °C sample was subjected to a three point flexural test in the same manner as the other samples. The 1000 °C sample was not tested as it fractured during handling before it could be tested. The single sample had a strength of 4.7 MPa, just more than a quarter of the 16.5 MPa of the original 7 wt % samples. However, it is important to note that the sample survived throughout the entire test (up to fracture) and did not crumble or fracture upon cooling or handling. Despite the drastically lower strength, the sample did survive the dehydration, which is an achievement in and of itself. Future experiments will focus on this aspect, testing not only for strength, but also determining the composition of the matrix after heating, and how it affects the bonding between the matrix and the reinforcement phase.

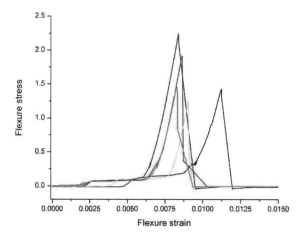

Fig. 1: Stress-strain curves for pure geopolymer bars. The average strength of the 5 samples was 1.7 MPa.

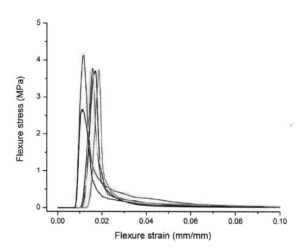

Fig. 2: Stress-strain curves for geopolymer bars reinforced with 1 wt % basalt fibers. The average strength of the 6 samples was 3.6 MPa.

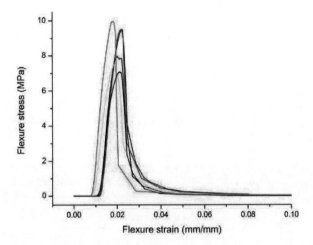

Fig. 3: Stress-strain curves for geopolymer bars reinforced with 3 wt % basalt fibers. The average strength of the 6 samples was 8.6 MPa.

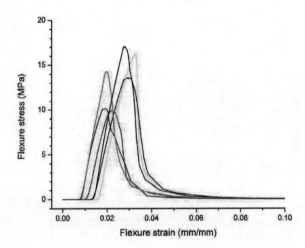

Figure 4: Stress-strain curves for geopolymer bars reinforced with 5 wt % basalt fibers. The average strength of the 6 samples was 13.5 MPa.

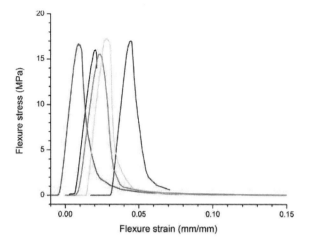

Fig. 5: Stress-strain curves for geopolymer bars reinforced with 7 wt % basalt fibers. The average strength of the 5 samples was 16.5 MPa.

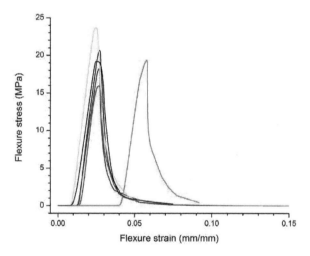

Fig. 6: Stress-strain curves for geopolymer bars reinforced with 10 wt % basalt fibers. The average strength of the 6 samples was 19.5 MPa.

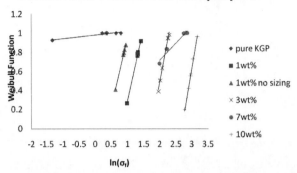

Fig. 7. Summary or Weibull statistics as a funtion of basalt fiber loading.

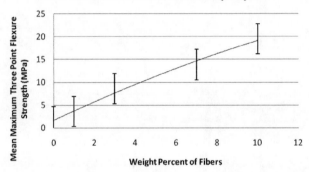

Fig. 8. Average 3-point flexure strength as a function of basalt fiber loading.

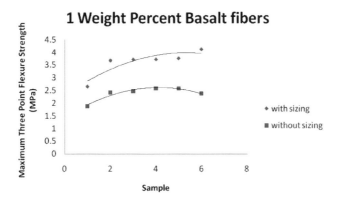

Fig. 9. Effect of the sizing on the three point bend strength of basalt composites.

The results of mechanical testing of chopped fiber reinforced composites are summarized in Table 1.

Table 1. Summary of average 3-point flexure strengths of chopped basalt fiber reinforced K-based geopolymer composites.

Wt % ¼" chopped basalt fiber	3-point flexure strength (MPa)
0	1.7
1	3.6
3	8.6
5	13.5
7	16.5
10	19.5

CONCLUSIONS

The addition of chopped basalt and polypropylene fibers significantly improved the room temperature bend strengths of potassium-based geopolymer composites. Basalt fibers of diameter 13 m x ¼" long yielded a 10-fold increase at 10 wt % loading. When the silane "sizing" was removed from the chopped fibers, the overall strength of the composite was significantly reduced. For example, even in 1 wt % fiber loadings the average strength was reduced from the 3.6 MPa to 2.4 MPa. The chopped basalt fibers significantly helped to retard cracking of geopolymer composites upon dehydration during heating to 500 °C for 1 hour. In one example, a single sample had a remnant strength of 4.7 MPa, after dehydration, which was approximately a quarter of the 16.5 MPa of the original 7 wt % chopped fiber reinforced samples.

ACKNOWLEDGEMENTS

This work was supported by a Grant from the United States Air Force Office of Scientific Research under Nanoinitiative Grant No. FA 9550-06-1-0221, through Dr. Joan Fuller. In addition, portions of this work were carried out in the Frederick Seitz Materials Research Laboratory Center for Microanalysis of Materials at the University of Illinois at Urbana-Champaign, which is partially supported by the U.S. Department of Energy under Grants DE-FG02-07ER46453 and DE-FG02-07ER46471. The authors thank Basaltex Inc, for supplying the samples of chopped basalt fibers.

REFERENCES

1. "Microstructure and Microchemistry of Fully-Reacted Geopolymers and Geopolymer Matrix Composites," W. M. Kriven, J. L. Bell and M.Gordon, Ceramic Transactions vol. **153**, 227-250 (2003).
2. "Composite Cold Ceramic Geopolymer in a Refractory Application," D. C. Comrie and W. M. Kriven, Ceramic Transactions vol. **153**, 211-225 (2003).
3. "Effect of Alkali Choice on Geopolymer Properties," W. M. Kriven and J. L. Bell, Cer. Eng. and Sci. Proc. vol. **25** [3-4] 99-104 (2004).
4. "Geopolymer Refractories for the Glass Manufacturing Industry," W. M. Kriven, J. L. Bell and M. Gordon, Cer. Eng. and Sci. Proc. vol. **25** [1] 57-79 (2004).
5. "Comparison of Naturally and Synthetically-Derived, Potassium-Based Geopolymers," M. Gordon, J. Bell and W. M. Kriven, Ceramic Transactions, vol. **165**. Advances in Ceramic Matrix Composites **X**, edited by J. P. Singh, N. P. Bansal and W. M. Kriven 95-106 (2005).
6. "Microstructural Characterization of Metakaolin-based Geopolymers," P. Duxson, G. C. Lukey, J. S. J. van Deventer, S. W. Mallicoat, W. M. Kriven Ceramic Transactions, vol. **165**. Advances in Ceramic Matrix Composites **X**, edited by J. P. Singh, N. P. Bansal and W. M. Kriven 71-85 (2005).
7. "Understanding the Relationship between Geopolymer Composition, Microstructure and Mechanical Properties," P. Duxson, J. L. Provis, G. C. Lukey, S. W. Mallicoat, W. M. Kriven and J. S. J. van Deventer, Colloids and Surfaces A – Physicochemical and Engineering Aspects, **269** [1-3] 47-58 (2005).
8. "Use of Geopolymeric Cements as a Refractory Adhesive for Metal and Ceramic Joins," J. L. Bell, M. Gordon and W. M. Kriven, Ceramic Engineering and Science Proceedings. Edited by D.-M. Zhu, K. Plucknett and W. M. Kriven, vol **26**, [3] 407-413 (2005).
9. "Novel Alkali-Bonded Ceramic Filtration Membranes," S. Mallicoat, P. Sarin and W. M. Kriven, Ceramic Engineering and Science Proceedings. Edited by M. E. Brito, P. Filip, C. Lewinsohn, A. Sayir, M. Opeka, W. M. Mullins, D.-M. Zhu and W. M. Kriven, vol **26**, [8] 37-44 (2005).
10. "Thermal Conversion and Microstructural Evaluation of Geopolymers or "Alkali Bonded Ceramics" (ABCs)," M. Gordon, J. Bell and W. M. Kriven. Ceramic Transactions, vol. **175**. Advances in Ceramic Matrix Composites **XI**. Edited by N.P. Bansal, J.P. Singh and W. M. Kriven, 225-236 (2005).
11. "Modeling Speciation in Highly Concentrated Alkaline Silicate Solutions," John L. Provis, Peter Duxson, Grant C. Lukey, Frances Separovic, Waltraud M. Kriven and

Jannie S. J. van Deventer, Industrial and Engineering Chemistry Research, **44** [23], 8899-8908 (2005).

12. "Geopolymers: More Than Just Cements," W. M. Kriven, J. Bell, M. Gordon and Gianguo Wen, pp 179-183 in Geopolymer, Green Chemistry and Sustainable Development Solutions, edited by Joseph Davidovits. Proc. World Congress Geopolymer, 2005, St. Quentin, France. Published by the Geopolymer Institute, St. Quentin, France (2005).

13. "Effect of Alkali and Si/Al Ratio on the Development of Mechanical Properties of Metakaolin-based Geopolymers," P. Duxson, S. W. Mallicoat G. C. Lukey, W. M. Kriven and J. S. J. van Deventer, Colloids and Surfaces A-Physicochemical and Engineering Aspects, **292**, 8-20 (2007).

14. "Intrinsic Microstructure and Properties of Metakaolin-Based Geoolymers," W. M. Kriven, J. L. Bell, S. W. Mallicoat and M. Gordon, contributed chapter to Proc. of Int. Worshop on Geopolymer Binders – Interdependence of Composition, Structure and Properties, Weimar, Germany, 71-86 (2007).

15. "Corrosion Protection Assessment of Concrete Reinforcing Bars with a Geopolymer Coating," W. M. Kriven, M. Gordon, B. L. Ervin and H. Reis. Cer. Eng. and Sci. Proc., **28** [9] 373-381 (2007). In Developments in Porous, Biological and Geopolymer Ceramics, edited by Manuel Brito, Eldon Case, W. M. Kriven, volume editors Jonathan Salem and Dongming Zhu, published by the American Ceramic Society.

16. "Laser Scanning Confocal Microscopy Analysis of Metakaolin-based Geopolymers," J. L. Bell, W. M. Kriven, A. P. R. Johnson, F. Caruso and J. S. J. van Deventer. Cer. Eng. Sci. Proc. **28** [9] 273-282 (2007). In Developments in Porous, Biological and Geopolymer Ceramics, edited by Manuel Brito, Eldon Case, W. M. Kriven, volume editors; Jonathan Salem and Dongming Zhu, series editors; published by the American Ceramic Society.

17. "Preparation of Ceramic Foams from Metakaolin-based Geopolymer Gels," J. L. Bell and W. M. Kriven. In Developments in Strategic Materials. Edited by Hua-Tay Lin, Kunihito Koumoto, Waltraud M. Kriven, David P. Norton, Edwin Garcia and Ivar Reimanis. Cer. Eng. Sci Proc. Vol **29** [10] 97-112 (2008).

18. "Atomic Structure of a Cesium Aluminosilicate Geopolymer: A Pair Distribution Function Study," J. L. Bell, P. Sarin, J. L. Provis, R. P. Haggerty, P. E. Driemeyer, P. J. Chupas, J. S. J. van Deventer and W. M. Kriven, Chemistry of Materials, **20** [14] 4768-4776 (2008).

19. "X-ray Pair Distribution Function Analysis of Potassium Based Geopolymer," J. L. Bell, P. Sarin, P. E. Driemeyer, R. P. Haggerty, P. J. Chupas and W. M. Kriven, Journal of Materials Chemistry, **18** [48], 5974 - 5981 (2008).

20. "Formation of Ceramics from Metakaolin-based Geopolymers: Part I. Cs-based Geopolymer," J. L. Bell, P. E. Driemeyer and W. M. Kriven, J. Amer. Ceram. Soc., **92** [1] 1-8 (2009).

21. "Formation of Ceramics from Metakaolin-based Geopolymers: Part II. K-based Geopolymer," J. L. Bell, P. E. Driemeyer and W. M. Kriven, J. Amer. Ceram. Soc., **92** [3] 607-615 (2009).

22. "Fabrication of Structural Leucite Glass-Ceramics from Potassium-based Geopolymer precursors," N. Xie, J. L. Bell and Waltraud M. Kriven, J. Amer. Ceram. Soc., **93** in press (2010).

NOVEL APPLICATIONS OF METAL-GEOPOLYMERS

Oleg Bortnovsky, Petr Bezucha
Research Institute of Inorganic Chemistry, Inc.
Revoluční 84, Ústí nad Labem, Czech Republic

Petr Sazama, Jiří Dědeček, Zdena Tvarůžková, Zdeněk Sobalík
J. Heyrovsky Institute of Physical Chemistry, Academy of Sciences of the Czech Republic,
Dolejškova 3, Prague, Czech Republic

Keywords: geopolymer, novel applications, metal cations, heterogeneous catalysis, antibacterial coating

ABSTRACT

Geopolymers are generally XRD-amorphous materials; however recent research has proved that they contain nanometer particles probably with zeolitic structure. It has also been confirmed in our investigation that sodium in metakaolin-based geopolymer could be easily exchanged for other cations such as ammonium, cobalt, copper etc. Moreover, according to UV-VIS spectra of Co-geopolymer, it has been proven that the local arrangements of Co^{2+}- extraframework ions in cationic positions correspond to those known in high-silica zeolites, such as mordenite, ZSM-5 or beta. A similar high level of ion exchange with various ions of transition metals has also been reached for metakaolin-slag-based geopolymers. Accordingly, it could be truly expected that these metal-exchanging geopolymers would show similar properties and potential applications as zeolites, while keeping the advantages of the geopolymers, such as simple synthesis procedure, easy molding into complicated shapes or forming thin highly adhesive layers. These novel applications, such as heterogeneous catalysis for environmental applications and active antibacterial thin coating of metal-geopolymers will be presented and discussed.

INTRODUCTION

The term geopolymer, introduced by Davidovits in 1979, represents inorganic materials usually prepared using reactions of various aluminosilicates with hydroxides or silicates of alkali and alkaline earth metals. In contrast to crystalline aluminosilicates, the synthesis of geopolymers proceeds at ambient or slightly elevated temperatures and atmospheric pressure and the whole volume of the synthetic mixture undergoes hardening. Alkaline activation with subsequent polymerization in the whole volume differentiates geopolymers from amorphous aluminosilicates. Due to the absence of long-range ordering, the amorphous structure of geopolymers is not well understood. Structural sequences formulated as –Al-O-Si-O-, -Si-O-Al-O-Si- and –Si-O-Al-O-Si-O-Si-O- were suggested to form a geopolymer network[1]. Recent progress in the analysis of geopolymers indicates that these materials should probably be regarded as amorphous analogues of zeolites[2,3]. Geopolymer materials exhibit a three dimensional network of SiO_4 and AlO_4 tetrahedra where the negative charge of the framework resulting from the presence of AlO_4 tetrahedra in the silicate framework is balanced by extra-framework cations, typically Na(I) or K(I). In analogy to zeolites SiO_4 and AlO_4 tetrahedra are arranged to form rings with variable size. It has been suggested that deformed six-, eight- and ten-membered rings are present in the geopolymer network. Such an arrangement results in properties similar to those of zeolites: (i) All extra-framework cations can be replaced by different ones (ion exchange)[3]. (ii) Bare cations can be located in a dehydrated geopolymer at cationic sites. It has been suggested that six or eight-membered rings accommodate bare divalent cations[3]. (iii) At least part of cations located at extra-framework sites is accessible for reactants[3]. As original Na or K-geopolymer was ion exchanged with a solution of an NH_4^+ salt followed by deammonization of a NH_4-geopolymer at elevated temperature or it was equilibrated with a diluted acid an acidic H-form of the geopolymer was finally obtained.

The close similarity of the geopolymer network and the zeolite framework in relation to ion exchange and accommodation of metal ions opens possibilities for the application of geopolymer materials as amorphous analogues of zeolites with applications for as adsorbents of cations from waste water was already reported[4]. However, the use of modified transition and noble metal-geopolymer materials in catalytic processes and silver-geopolymer for antibacterial applications has not been discussed.

The present work is concerned with the synthesis and application of new types of metal-geopolymer materials for industrially important, heterogeneously catalysed reactions and geopolymer coating with antibacterial properties. Two heterogeneously catalysed reactions were selected to demonstrate the catalytic properties of transition/noble metal-geopolymer catalysts: (i) selective catalytic reduction of nitrogen oxides by ammonia, (ii) catalytic oxidation of VOCs and geopolymer coating was chosen for demonstration of antibacterial properties.

Generally, oxidic supported catalysts such as V_2O_5/Al_2O_3, V_2O_5/TiO_2 or V_2O_5-WO_3/TiO_2 and developed catalysts based on Cu- or Fe-zeolites[5,6] in form of extrudates or honeycomb were used for selective catalytic reduction of nitrogen oxides by ammonia and noble metal catalysts on various supports such as alumina, silica, etc. were used for catalytic oxidation of VOCs[7]. Potential application of geopolymer catalysts for these processes with simply preparation procedure, high robustness, high thermal and thermal-shock resistance and low production costs could be very interesting.

Antibacterial inorganic coating is nowadays generally prepared from various silver-containing powder alumosilicates with the addition of particularly organic-based binder[8] or as glass-like materials[9] treated at temperatures above 400 °C. Application of geopolymer coating prepared at temperatures as low as room one or below 100 °C together with reported[10] high adhesion at majority of inorganic surfaces makes the method very promising.

Both catalytic and antibacterial applications are patent pending in Czech Republic.

EXPERIMENTAL

Preparation of geopolymer catalysts and antibacterial coating

Table I. lists the parent geopolymers geopolymer catalysts and geopolymer coatings used in this study. Parent Na-Geo-1 geopolymer with sodium as a charge compensating cation, was prepared by mixing of 10 g metakaolin (43.5 wt.% of Al_2O_3 and 53.7 wt.% of SiO_2) with an alkali activator containing 12.2 g of water glass (31.95 wt.% of SiO_2, 17.73 wt.% of Na_2O; molar ratio SiO_2/Na_2O of 1.86) and 1.1 g of NaOH (49.38 wt.% in water) and 0.7 g of water. Polymerization was performed in a closed vessel at 60 °C for 48 hours. After maturing of the geopolymer for one week, the geopolymeric monolith was ground. The particle size of the powdered geopolymer ranged between 5 and 50 µm.

K,Ca-Geo-2 sample with potassium as a charge-compensating cation was prepared by mixing of 10 g metakaolin (43.5 wt.% of Al_2O_3 and 53.7 wt.% of SiO_2) and 8 g of grinding granulated blast furnace slag with 16.2 g of potassium water glass (18.3 wt.% of SiO_2, 17.3 wt.% of K_2O; molar ratio SiO_2/K_2O of 1.66). Polymerization was performed in a closed vessel at RT. After maturing of the geopolymer for four weeks, the geopolymeric monolith was ground. The particle size of the powdered geopolymer ranged between 5 and 50 µm.

K-Geo-3 coating sample with potassium as a charge-compensating cation was prepared by mixing of 5 g metakaolin (42.8 wt.% of Al_2O_3 and 54.5 wt.% of SiO_2) and 10 g of thermal silica (93.8 % wt. SiO_2 a 2.9 % wt. Al_2O_3) with 11.8 g of potassium water glass (15.56 wt.% of SiO_2, 24.39 wt.% of K_2O; molar ratio SiO_2/K_2O of 1.00). Geopolymer resin was spread on a degreased microscopic glass in a 30 μm thick layer. Sample was left in an open air for 24 hours and then dried at 85 °C for 3 days.

AgK-Geo-3 antibacterial coating sample was prepared by the similar way as K-Geo-3 sample, but 0.8 ml of 0.1 M solution of $AgNO_3$ was added at the end of resin preparation, accompanied with instant darkening of resin. Silver content in resulting AgK-Geo-3 sample was 170 ppm.

After immersion for 24 hours in distilled water and drying at RT, K-Geo-3 coating sample was repeatedly sprayed with 0.36 M solution of $AgNO_3$. The coated sample was finally dried at RT for 72 hours at daylight, while the sample darkened again. Silver content in resulting KAg -Geo-3 sample was 450 ppm.

Table I. List of used geopolymer based catalyst and coating

Sample	Preparation method	Catalytic reaction/ antibacterial coating
Na-Geo-1	Metakaolin-based Na-geopolymer	Parent geopolymer
KCa-Geo-2	Metakaolin-slag-based K-geopolymer	Parent geopolymer
K-Geo-3	Metakaolin-based K-geopolymer	Parent geopolymer coating
NH_4-Geo-1	Ion exchange of Na-Geo-1 with NH_4NO_3 solution	Selective catalytic reduction of NO_x by NH_3
NH_4-Geo-2	Ion exchange of KCa-Geo-2 with NH_4NO_3 solution	Parent NH_4-form geopolymer
$CuNH_4$-Geo-1	Ion exchange of NH_4-Geo-1 with $Cu(NO_3)_2$ solution	Selective catalytic reduction of NO_x by NH_3
$CuNH_4$-Geo-2	Ion exchange of NH_4-Geo-2 with $Cu(NO_3)_2$ solution	Selective catalytic reduction of NO_x by NH_3
$PtNH_4$-Geo-2	Ion exchange of NH_4-Geo-2 with $Pt[(NH_3)_4Cl_2]$ solution	Total oxidation of decane as VOC
FeKCa-Geo-2	Impregnation of Kca-Geo-2	Total oxidation of decane as VOC
$CoNH_4$-Geo-2	Ion exchange of NH_4-Geo-2 with $Co(NO_3)_2$ solution	Total oxidation of decane as VOC
AgK-Geo-3	$AgNO_3$ in geopolymer resin	Antibacterial coating
KAg-Geo-3	Spraying of $AgNO_3$ onto K-Geo-3	Antibacterial coating

The NH_4^+ form of the Geo-1 and Geo-2 geopolymers was obtained by ion-exchange procedure with an aqueous 0.5 M NH_4NO_3 solution at RT (100 ml of solution per 1 g of geopolymer applied thrice over 12 h). After each ion exchange procedure, the geopolymers were carefully washed with distilled water and dried in the open air.

$CuNH_4$-Geo-1 and $CuNH_4$-Geo-2 geopolymer catalysts were prepared by Cu(II) ion exchange of the corresponding parent geopolymer in the NH_4^+ cationic form with aqueous 0.1 M $Cu(NO_3)_2$ solution at RT for 24 h. After ion-exchange, the material was filtered, washed with H_2O and dried in the open air at RT.

CoNH$_4$-Geo-2 geopolymer catalysts were prepared by Co(II) ion exchange of the corresponding parent geopolymer in the NH$_4^+$ cationic form with aqueous 0.1 M Co(NO$_3$)$_2$ solution at RT for 24 h. After ion-exchange, the material was filtered, washed with H$_2$O and dried in the open air at RT.

FeKCa-Geo-2 geopolymer catalyst was prepared using an impregnation procedure with a solution of FeCl$_3$ in acetyl acetone. The procedure is currently used for preparation of Fe-zeolites yielding highly dispersed iron species. Details of the procedure are described in Ref 11.

PtNH$_4$-Geo-2 catalyst was obtained by adding 100 ml of 2 % Pt[(NH$_3$)$_4$Cl$_2$] aqueous solution to a suspension of 1 g NH$_4$-Geo-2 in 100 ml H$_2$O. The mixture was stirred for 24 h, filtered, dried in the open air at RT and calcinated in an air stream at 450°C.

The molar ratios of the major framework elements and metal cations in parent geopolymers, geopolymer catalysts and coatings determined by XFS are listed in Table II.

Table II Molar ratios of main elements and cations active sites in parent geopolymers, catalysis and coating estimated by XRF

Sample	Si/Al	Na(K)/Al	Ca/Al	Cu/Al	Co/Al	Fe/Al	Pt/Al	Ag ppm
Na-Geo-1	1.88	1.10	0.01	-	-	0.008	-	-
KCa-Geo-2	1.82	0.68	0.84	-	-	0.024	-	-
K-Geo-3	4.84	1.38	0.003	-	-	0.024	-	-
NH$_4$-Geo-1	1.87	0.25	0.01	-	-	0.005	-	-
NH$_4$-Geo-2	1.85	0.25	0.72	-	-	0.025	-	-
CuNH$_4$-Geo-1	1.88	0.02	0.01	0.33	-	0.005	-	-
CuNH$_4$-Geo-2	1.84	0.07	0.18	0.67	-	0.029	-	-
PtNH$_4$-Geo-2	1.95	0.19	0.70	-	-	0.024	0.026	-
FeKCa-Geo-2	1.94	0.74	0.91	-	-	0.073	-	-
CoNH$_4$-Geo-2	1.87	0.14	0.13	-	0.16	0.026	-	-
AgK-Geo-3	4.85	1.39	0.004	-	-	0.023	-	170
KAg-Geo-3	4.84	1.38	0.003	-	-	0.024	-	450

Structural analysis of geopolymer catalysts and coating

The amorphous nature of the samples was checked by X-ray powder diffraction analysis on a Philips MPD 1880 diffractometer with CuKα radiation in the Bragg-Brenato geometry and equipped with a graphite monochromator and scintillation counter. The X-ray pattern was measured in the region of 5 – 75 ° of 2 theta with a 0.04 ° step.

The ATR FTIR spectroscopy technique was used to monitor the degree of polymerization of the geopolymers. The spectra were recorded on a Nicolet Protégé 460 FTIR spectrometer equipped with an MCT/A detector.

Transformation of metakaolin to a geopolymer was monitored by ^{27}Al MAS NMR spectroscopy. ^{27}Al MAS NMR experiments were carried out using a Bruker Avance 500 MHz (11.7 T) Wide Bore spectrometer with 4 mm o.d. ZrO$_2$ rotors with a rotation speed of 12 kHz. High-power decoupling pulse sequences with /12 (0.7 μs) excitation pulse were employed to collect ^{27}Al MAS NMR single pulse spectra. The ^{27}Al NMR observed chemical shift was referred to an aqueous solution of Al(NO$_3$)$_3$.

Nitrogen sorption isotherms of parent geopolymers were measured with a ASAP 2020 (Micromeritics Instrument Corporation, Norcross, USA) volumetric instrument at − 196 °C. Prior to the sorption measurements, all the samples were degassed at 250 °C for at least 24 hours until a pressure of 10^{-3} Pa was attained.

The chemical compositions of parent geopolymers and geopolymers catalysts were determined by XFS with a Philips PW 1404 spectrometer, equipped with the UniQuant analytic program yielding semi-quantitative analysis of 74 elements from fluorine to uranium with 10 % standard deviation.

Catalytic experiments

Selective catalytic reduction of NO_x by NH_3

A fixed-bed flow-through quartz microreactor was used to analyse the catalyst activity of geopolymer catalysts in NH_3-SCR-NO_x. The composition of gases at the inlet of the reactor was typically 500 ppm NO and 500 NO_2, 2.5 O_2, and 1000 ppm NH_3 and He as a carrier gas. The weight of the catalyst (0.3 – 0.5 mm grains) was 50 mg, and the flow rate was 350 cm^3 min^{-1}, corresponding to a GHSV of 210 000 h^{-1}. All the gas connections were kept at a temperature of 200 °C to avoid the formation and deposition of NH_4NO_3. The concentrations of NO, NO_2 and NH_3 were monitored with a NO/NO_x/NH_3 analyzer (ABB Limas 11). The concentrations of N_2 and N_2O were determined by an on-line connected Hewlett-Packard 6890 gas chromatograph. Columns of Poraplot Q and Molecular sieve 5A were used for the separation and TCD for the detection.

Total oxidation of hydrocarbons

Partial pressure saturators and mass flow controllers were used for feeding decane vapors and oxygen into a stream of He. The feed contained 250 ppm of n-decane, 6 % O_2 and He as a carrier gas. An amount of 100 mg of catalysts of particle size of 0.3 – 0.5 mm and total flow rate of 100 ml.min^{-1} corresponded to a GHSV of 60,000 h^{-1}. The concentrations of O_2, CO, CO_2 and hydrocarbons were determined by an on-line connected Hewlett-Packard 6890 gas chromatograph. Two gaseous samples were injected with two 10-port valves. Columns of Poraplot Q and Molecular sieve 5A and a thermal conductivity detector (TCD) were used to separate and detect O_2, CO and CO_2. A Poraplot Q column separated CO_2 from O_2 and traces of N_2 and CO, and a six-port valve was used to bypass the column with Molecular sieve 5A during the analysis of CO_2. In the second branch, hydrocarbons were separated on a Poraplot Q column and detect by FID.

Evaluation of antibacterial efficiency

Extinction of cultivated micro organisms at 37 °C by samples of geopolymer coating on a microscopic glass was tested by immersion into 200 ml of day water with a concentration of 8.0×10^3 CFU /1ml (CFU: colony forming unit) for 24 and 48 hours according to Czech/EU standard ČSN EN ISO 6222. A blank test and test with a coating sample without silver was done for comparison of an efficiency of geopolymer coating. Leachability of silver from geopolymer coating in day water was tested by immersion for 48 hours in independent experiments, and silver content in the extract was determined by ICP technique.

RESULTS AND DISCUSSION

Structure of geopolymers

Surface area calculated by the BET method, pore volume obtained by BJH and t-plot methods and the pore diameters of the synthesised parent geopolymers are summarised in Table III. It is shown that the synthesis yielded mesoporous parent geopolymers with mesopore volume of 0.089 and 0.095 $cm^3.g^{-1}$ for Na-Geo-1 and KCa-Geo-2, respectively, and negligible microporosity (see Figure 1 and Table III). It is necessary to add here that especially in case of metakaolin-slag based geopolymer (KCa-Geo-2) silicate module of alkali activator could greatly influenced mesoporosity of resulted geopolymer. The sample of Davya 60 with alkali activator (hardener) with silicate module approx. 2.0, prepared in our laboratory from raw materials obtained from Cordi-Geopolymere, demonstrated almost four time higher mesopore volume and almost five time higher BET surface area (see Table III). In the case of Davya 30 and Davya 65 prepared with potassium and sodium alkali activator with much lower silicate modulus approx. from 1.25 to 1.45 the obtained textural properties were close to KCa-Geo-2 sample used for catalyst preparation. The observed mesoporosity is in good agreement with the results reported by Kriven[2], who observed that the geopolymer material consists of nanoparticulates ranging from 5 to 15 nm, separated by nanopores of 3 to 10 nm. The maximum range of the pore size (10 and 8 nm for Na-Geo-1 and KCa-Geo-2) indicates that geopolymer-based catalysts could accommodate spacious reactants and intermediates. The surface area is high enough to accommodate a number of active sites for the catalytic performance. In contrast to the standard synthesis of geopolymer materials, which leads to the formation of compact materials of low porosity and surface area, the procedures employed in this study yield material with suitable textural properties. Moreover, it is demonstrated that materials with lower price and a very simple procedure can be used for geopolymer synthesis.

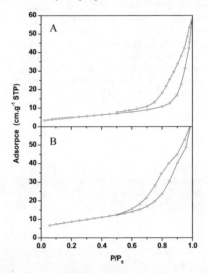

Figure 1. Nitrogen adsorption isotherms of Na-Geo-1 (A) and KCa-Geo-2 (B).

Table III. Textural properties of parent geopolymers and Davya geopolymers for comparison.

Sample	$S_{BET}{}^a$ $(m^2 . g^{-1})$	$V_{BJH}{}^b$ $(cm^3 . g^{-1})$	$V_{Mi}{}^c$ $(cm^3 . g^{-1})$	D^d (nm)	$D_{max}{}^e$ (nm)
Na-Geo-1	17.1	0.089	0.0032	8-12	10
KCa-Geo-2	31.6	0.095	0	6-10	8
Davya 60	140.3	0.358	0.0289	5-9	6
Davya 30	38.2	0.094	0.0074	5-9	7
Davya 65	32.2	0.165	0.0067	10-30	22

$^a S_{BET}$ – surface area
$^b V_{BJH}$ – mezopore volume, calculated by the BJH method
$^c V_{Mi}$ - micropore volume determined by the t-plot method
$^d D$ – pore size range
$^e D_{max}$ – maximum of the pore size range

The XRD patterns of the powdered Na-Geo-1 and KCa-Geo-2 samples are depicted in Figure 2. The broad band around $2\theta = 30°$ indicates that the prepared samples are amorphous and contain only traces of crystalline compounds. This is reflected in the XRD pattern of Na-Geo-1 kaolin by a weak sharp signal at $2\theta = 19.8$, quartz is manifested by sharp signals at $2\theta = 26.6$ °, mullite by weak sharp signals at $2\theta = 16.4, 25.4, 35.2, 39.4, 40.9, 42.7$ and $60.7°$ and magnetite by a weak signal at $2\theta = 33.2°$. All these crystalline impurities originated from the metakaolin used for preparation of both geopolymer samples. In the XRD pattern of KCa-Geo-2, kaolin is manifested by a weak signal at $2\theta = 19.8$, quartz by weak signals at $2\theta = 26.6°$, calcite by sharp signals at $2\theta = 29.4$ and merwinite $(Ca_3Mg(SiO_4))$, by a number of sharp signals at $2\theta = 32.5, 33.4, 33,8, 39.5, 40.8, 41.7, 43.3, 44.6, 47.6, 48.5$ and $60.4°$. Merwinite and calcite are crystalline parts of granulated blast furnace slag used for preparation of the KCa-Geo-2 geopolymer.

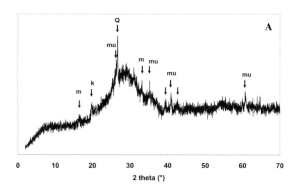

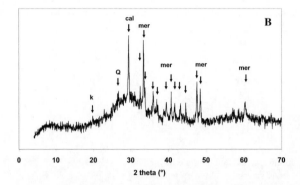

Figure 2. X-ray diffraction pattern of Na-Geo-1 (A), KCa-Geo-2 (B). k, kaolin; Q, quartz; mu, mullite; m, magnetite, cal, calcite;.mer, merwinite.

Figure 3 depicts the FTIR spectra of NH_4-Geo-1 and NH_4-Geo-2 geopolymers in the region of the T-O-T vibration. Three absorption bands typical of the geopolymer material were observed in the spectra at 870, 995 and 1120 cm^{-1}. The bands with maxima at 995 cm^{-1} (low frequency – LF band) and 1120 cm^{-1} (high frequency – HF band) are ascribed to the antisymmetric T–O–T vibration[3,12-15].

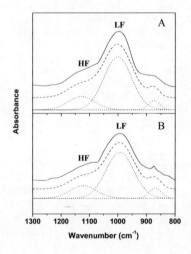

Figure 3. ATR FTIR spectra of NH_4-Geo-1 (A) and NH_4-Geo-2 (B) geopolymers. Experimental spectrum (—), simulated spectrum (---) and its decomposition to individual Gaussian bands (....).

The ratio of the low and high frequency bands (LF/HF) can be used for rough estimation of the degree of polymerization of the geopolymer network[15]. The LF/HF ratios of 4.1 and 3.7 for NH_4-Geo-1 and NH_4-Geo-2, respectively, indicate a high degree of polymerization. The band at 873 cm^{-1} reflects the Si-OH bending vibration[12].

The ^{27}Al MAS NMR spectra of the Na-Geo-1 and KCa-Geo-2 samples are depicted in Figure 4 and compared with the spectrum of the metakaolin used for their preparation. The resonance at 58 ppm corresponds to tetrahedrally coordinated Al atom and dominates spectra of both geopolymers. The resonance at 3 ppm can be ascribed to octahedrally coordinated Al of residual unreacted metakaolin and represents less than 4 % of all the Al atoms. This indicates nearly complete transformation of metakaolin into geopolymer. In the case of K-Geo-3 geopolymer the spectrum is more complicated and demonstrated at least four resonances at 4.5, 15, 30 and 53 ppm, known for metakaolin and unreacted kaolin. It evidenced that in the case of geopolymer prepared with two solid aluminosilicate components with different reactivity thermal silica and metakaolin prepared from shale calcined at 750 °C the later remain partly unreacted, and do not incorporate into geopolymer network and remained as microfiller. Therefore Si/Al ratio in geopolymer phase itself is much higher than was estimated by XRF. The similar effect was observed early in phosphorus containing matrices used for fiber reinforced composites[16]. Usage of thermal silica for preparation of geopolymer coating is necessary to achieve a good adhesion with majority of inorganic surfaces[10].

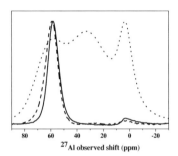 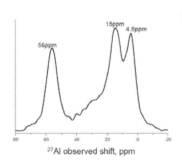

Figure 4. ^{27}Al MAS NMR spectra of Na-Geo-1 (——), KCa-Geo-2 (- - -) and metakaolin used for geopolymer synthesis (••••) (left) and AgK-Geo-3 (right).

3.2   Active structures in geopolymer catalysts

Transition metals and platinum as catalytic components, which have been proven to provide high catalytic activity in commercially used catalysts for catalytic reduction of nitrogen oxides by ammonia, catalytic oxidation of VOC, were incorporated into Na-Geo-1 and KCa-Geo-2 geopolymers. Divalent Cu(II) ions, which exhibit high activity and selectivity in NH_3-SCR-NO_x, were introduced into Na-Geo-1 and KCa-Geo-2 by a common ion-exchange procedure. It has recently been reported[3] that the internal volume of an originally sodium geopolymer can easily communicate with the external environment of the geopolymer and thus that all extra-framework cations balancing the negative charge of the geopolymer network can be ion-exchanged and at least some of them can be accessible for reactants. It is clear from the chemical analysis (see Table II) that, using ion exchange procedures, almost 98 % of the Na^+ was exchanged with Cu^{2+} and NH_4^+ ions at CuNH4-Geo-1 and 80 % and 90 % of Ca^{2+} and K^+, resp., were exchanged at CuNH4-Geo-2. It is noticeable that the exchange of divalent Ca^{2+} was much more prominent only with divalent Cu^{2+} and Co^{2+} cations in comparison with univalent NH_4^+ ions. This indicates that the ion-exchange procedure of geopolymers provided predominantly Cu^{2+} ions located in the dehydrated geopolymer at cationic sites.

Platinum, iron and cobalt are typical components of catalysts based on supported noble metals or oxides of transition metals for total oxidation of VOC[17,18]. The common ion-exchange procedure employed to prepare CoNH4-Geo-2 led analogously to CuNH4-Geo-2 with ion-exchange of 80 % and 85 % of Ca^{2+} and K^+, resp., by Co^{2+} and NH_4^+ ions, as indicated by the chemical analysis. Extra-network Co^{2+} ions balancing network Al negative charge in the dehydrated Co-geopolymer exhibit UV-Vis spectra, which indicate the presence of three types of well-defined local structures accommodating Co^{2+} ions. Structural analysis of the Co species in geopolymer materials has been published elsewhere[3].

The ion-exchange of NH_4^+ with Fe-ions and limited rate of the formation of Fe-oxo species have been already found for the method employing $FeCl_3$ + ACAC mixture for impregnation of zeolites[19]. Thus, formation of highly dispersed iron species is expected for the FeKCa-Geo-2 catalyst. However, analysis of Fe species incorporated to geopolymers using UV-Vis spectroscopy is significantly limited due to the presence of Fe oxides in metakaolin used for geopolymer preparation, cf. chemical analysis of prepared geopolymers.

Impregnation with $Pt[(NH_3)_4Cl_2]$ solution with subsequent calcination should provide clusters of platinum species of undefined sizes.

The similar silver clusters could be formed at two different methods of preparation of Ag-geopolymer antibacterial coating. As followed from evaluation of Ag leaching some Ag cations were also presented, mostly in case of spraying of sample which is close to ion exchange procedure

Analysis of these structures is beyond the scope of this paper and requires further analysis.

Activity of geopolymer catalysts in NH_3-SCR-NO_x

The NO_x and NH_3 conversions in the NH_3-SCR-NO_x reaction as a function of reaction temperature for Cu-geopolymers, NH4-geopolymer and commercial V_2O_5/Al_2O_3 catalyst are compared in Figure 5. For NH4-Geo-1, the NO_x and NH_3 conversion began at 300 °C and increased with increasing reaction temperature. The activity is essentially related to the presence of acid sites important for adsorption of basic molecules of ammonia and traces of transition metals involved in the redox cycle. The introduction of copper into the NH4-Geo-1 geopolymer resulted in markedly improved catalyst activity over the entire temperature range. CuNH4-Geo-1 exhibits significant conversion of NO_x already at 250 °C, which increases with the reaction temperature, passing through a maximum at 400°C and then declining slightly. The important observation is that NH_3 is completely oxidized already at a temperature of 230 °C and no NH_3 slip is present above this temperature. The conversion of NO_x was already observed for CuNH4-Geo-2 containing a high amount of copper at temperatures below 200 °C; however, then the conversion of NO_x declined at temperatures above 250 °C due to the lack of non-selective oxidisation of NH_3 by O_2, as is indicated by the steadily increasing conversion of NH_3. The obtained conversion of NO_x and NH_3 on geopolymer catalysts is compared with a commercial sample of commonly used V_2O_5/Al_2O_3 catalyst. V_2O_5/Al_2O_3 exhibits slightly increasing conversion of NO_x and NH_3 with increasing temperature and highly selective utilization of NH_3. The high selectivity resulted in a high ammonia slip. From comparison of conversions of NO_x obtained for geopolymer and the commercial catalysts, it is clear that CuNH4-Geo-1 could compete with the common catalyst in the temperatures range of 250 - 350°C. The higher conversion of NH_3 over CuNH4-Geo-1 compared to the V_2O_5/Al_2O_3 catalyst can be advantageous in processes with limited possibility of precise dosing of NH_3 into the deNOx process; however, it can slightly decrease utilization of NH_3 under steady-state conditions.

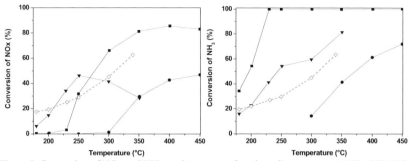

Figure 5. Conversion of NOx and NH_3 to nitrogen as a function of temperature at NH_3-SCR-NOx over (●) NH_4-Geo-1, (■) NH_4Cu-Geo-1, NH_4Cu-Geo-2 (▼) and V_2O_5/Al_2O_3 (◊) catalysts. Conditions: 500 ppm NO, 500 ppm NO_2, 1000 ppm NH_3 and 2.5 % O_2 in a He stream, GHSV = 210 000 h⁻¹.

Total oxidation of VOC

Conversion of decane and selectivity for CO_2 and CO in the total oxidation of decane as a function of temperature over $PtNH_4$-Geo-2, $FeKCa$-Geo-2 and $CoNH_4$-Geo-2 catalysts is shown in Figure 6. The reactions were carried out at an oxygen concentration of 6 %, i.e. lower oxygen concentration than in the air, which is typical for exhaust gases of combustion processes. $PtNH_4$-Geo-2 already exhibits high conversion at very low temperatures with 50 % conversion of decane obtained at 135°C. The oxidation is highly selective for CO_2 whereas CO has not been detected in the products of combustion even at temperatures of approx. 100°C and oxygen concentration of 6 %. The working temperature window for both $FeKCa$-Geo-2 and $CoNH_4$-Geo-2 catalysts is understandably shifted to higher temperatures than for the Pt-geopolymer. The reaction starts at 260°C and total conversion is achieved at 470 - 480°C. The selectivity for CO_2 exhibits an increasing tendency with increasing temperature when CO is completely oxidised to CO_2 at 470°C over $CoNH_4$-Geo-2.

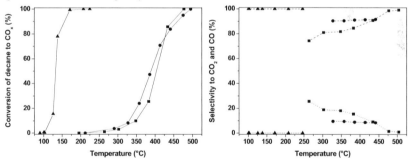

Figure 6. Conversion of decane (solid line) and selectivity for CO_2 (dash line) and CO (dot line) at total oxidation as a function of the temperature over $PtNH_4$-Geo-2 (▲), $FeKCa$-Geo-2 (●) and $CoNH_4$-Geo-2 (■). GHSV = 60 000 h⁻¹, 250 ppm decane and 6 % O_2 in a He stream.

Decane is one of the major components of diesel fuel, and automotive transport represents one of the most important sources of VOC pollution[20]. Thus the high conversions achieved with decane at low and medium temperatures, together with the high thermal stability of the geopolymers, exemplified the possible potential for utilization of geopolymer catalysts in total oxidation of VOC.

Antibacterial activity of geopolymer coating

Antibacterial efficiency of Ag-containing geopolymer coating, of parent geopolymer coating, and concentration of leached Ag in water estimated by ICP in independent experiments are compared in Tab. IV. It is clearly seen that for all types of silver-containing geopolymer coating almost 100 % antibacterial efficiency was attained within 24 hours of exposition. It is generally known that geopolymer itself, due to increased surface alkalinity, is not damaged by biological factors. Nevertheless, this effect is not sufficient enough to extinct microorganisms in water, which is in contact with, as can be seen from Table IV. The results of leacheability of Ag into water further evidenced, that in the case of postsynthesis treatment of geopolymer by spraying with a solution of silver nitrate the concentration of free Ag-ions in the treated water is too high (0.16 mg/l) and exceed, i.e. the limit for drinking water (0.05 mg/l). In the case of geopolymer coating with silver nitrate added directly to geopolymer resin, however, concentration of free Ag ions is almost 10 times lower (0.02 mg/l) and it is under the limit for drinking water.

Table IV. Antibacterial efficiency of geopolymer coating and concentration of free leached Ag

Sample	estimated (CFU/1 ml)		free Ag-ions in water, mg/l
	24 hrs	48 hrs	
AgK-Geo-3	10	0	0.02
KAg-Geo-3	0	0	0.16
K-Geo-3	1.65×10^4	1.0×10^4	-
blank	0.8×10^4	-	-
limit in drinking water	-	-	0.05

CONCLUSIONS

Geopolymer-based alumosilicate materials represent a new type of redox heterogeneous catalysts. Transition metal ions and noble metal ions located in extra-network positions of the geopolymer can be introduced by an ion-exchange procedure. Functionalized geopolymers possess catalytic properties in heterogeneously catalysed redox reactions. The redox catalytic properties of the Pt, Fe, Cu and Co sites in geopolymer catalysts were demonstrated on NH_3-SCR-NO_x by ammonia and total oxidation of volatile hydrocarbons. In some aspects, the catalytic properties of the prepared catalysts can approach materials commonly used in industrial installations. The geopolymer catalysts exceed currently used catalytic materials in simplicity of preparation and robustness.

Geopolymer-based silver containing inorganic coating represent a novel type of simply prepared antibacterial coating with combination of high antibacterial activity of negligible amount of added silver incorporated in geopolymer network and high adhesion to majority of inorganic surface, i.e. concrete, bricks, metal etc.

It can be concluded that, due to their high thermal resistance, low cost production, easy fabrication and formation/accommodation of redox and antibacterial active centres, geopolymer-based materials have a potential for field application of catalysts and for germicide. Successful application of geopolymers in these areas requires, however, detailed understanding of these materials, their structure and the relationship to various aspects of their properties.

REFERENCES

[1] J. Davidovits, Geopolymers: Inorganic Polymeric New Materials, *Journal of Thermal Analysis,* **37** 1633-1656 (1991).

[2] W.M. Kriven, J.L. Bell, M. Gordon, Microstructure and Michrochemistry of Fully-Reacted Geopolymers and Geopolymer Matrix Composites, *Ceramic Transactions,* **153** 227 (2003).

[3] O. Bortnovsky, J. Dedeček, Z. Tvarůžková, Z. Sobalík, J. Šubrt, Metal Ions as Probes for Characterization of Geopolymer Materials, *Journal of the American Ceramic Society* **91** 3052-3057 (2008).

[4] L. Li, S.B. Wang, Z.H. Zhu, Geopolymeric Adsorbents from Fly Ash for Dye Removal from Aqueous Solution, *Journal of Colloid and Interface Science,* **300** 52-59 (2006).

[5] Z.Sobalík, B. Wichterlová, M. Markvart, I. Jirka, A. Smiesková, Process for Preparing a Catalyst for Selective Catalytic Reduction of Nitrogen Oxides Based on Cu-Zeolites, patent CZ284749 (1992).

[6] W. Hoelderich, O. Scheidsteger, R. Drews, W. D. Mross, K. Hess, H. Schachner, Process for eliminating nitrogen oxide from exhaust gases, patent EP0299294 (1989).

[7] J.M. Chen, P. H. Nguyen, Catalytic Oxidation Catalyst and Metod for Controlling VOC, CO and Halogenated Organic Emission, US patent 5578283 (1996).

[8] M. Mizugzchi, T.Yanase, Antibacterial porous Inorganic Capsule and Production Thereof, patent JP6134290 (1994).

[9] K. Nakada, Antibacterial Glass and Resin Composition Containing the Same, patent JP2001226139 (2001).

[10] J.Davidovits, Method for Bonding Fiber Reinforcement on Concrete and Steel Structures and Resultant Products, US Patent 5925449 (1999).

[11] N.K. Sathu, P. Sazama, V. Valtchev, B. Bernauer, Z. Sobalik, Oxidative Dehydrogenation of Propane over Fe-BEA Catalysts, *Studies in Surface Science and Catalysis,* **Vol. 174, part 2**, 1127-1130, (2008).

[12] V.F.F. Barbosa, K.J.D. MacKenzie, C. Thaumaturgo, Synthesis and Characterisation of Materials Based on Inorganic Polymers of Alumina and Silica: Sodium Polysialate Polymers, *International Journal of Inorganic Materials,* **2** 309-317 (2000).

[13] J.L. Provis, G.C. Lukey, J.S.J. Van Deventer, Do geopolymers actually contain nanocrystalline zeolites? A re-examination of existing results, *Chemistry of Materials,* **17** 3075-3085, (2005).

[14] G. Coudurier, C. Naccache, J.C. Vedrine, Uses of I.R. Spectroscopy in Identifying ZSM Zeolite Structure, *Journal of the Chemical Society, Chemical Communications,* 1413-1415 (1982).

[15] O. Bortnovsky, Z. Sobalík, Z. Tvarůžková, J. Dědeček, P. Roubíček, Ž. Průdková, M. Svoboda, Structure and Stability of Geopolymers synthesized from kaolinitic and shale residues, *Proceedings of Geopolymer 2005 World Congress. Geopolymers, Green Chemistry and Sustainable Development Solutions, Edited by J. Davidovits. Saint-Quentin, France,* 81-84 (2005).

[16] O. Bortnovsky, P. Bezucha, J. Dědeček, Z. Sobalík, V. Vodičková, D. Kroisová, P. Roubíček, M. Urbanová, Properties of Phosphorus-Containing Geopolymer Matrix and Fiber- Reinforced Composite, *Ceramic Engineering and Science Proceedings,* **Vol. 30, Issue 2** 283-300 (2009).

[17] S.C. Kim, The Catalytic Oxidation of Aromatic Hydrocarbons over Supported Metal Oxide, *Journal of Hazardous Materials,* **91** 285-299 (2002).

[18] J.M. Giraudon, A. Elhachimi, G. Leclercq, Catalytic Oxidation of Chlorobenzene over Pd/Perovskites, *Applied Catalysis B: Environmental,* **84** 251-261 (2008).

[19] Z. Sobalík, B. Wichetrlova, M. Markvart, Z. Tvaruzkova, Process for preparing a zeolite-based catalyst for removing nitrogen oxides from exhaust gases by reduction with hydrocarbons, patent CZ 293917 (2004).
[20] C. Badol, N. Locoge, J.C. Galloo, Using a Source-Receptor Approach to Characterise VOC Behaviour in a French Urban Area Influenced by Industrial Emissions: Part II: Source Contribution Assessment Using the Chemical Mass Balance (CMB) model, *Science of the Total Environment*, **389** 429-440 (2008).

ACKNOWLEDGEMENT

This work was supported by the Ministry of Industry and Trade of the Czech Republic (# FT-TA4/068).

MAKING FOAMED CONCRETES FROM FLY ASH BASED ON GEOPOLYMER METHOD

Nhi Tuan Pham and Hoang Huy Le
HCM City Institute of Resources Geography
HoChiMinh City, Vietnam

ABSTRACT

The foamed concretes which have useful characteristics for modern technique and constructions have been widely interested. The article presents the first results on the manufacture of foamed concretes from fly ash based on Geopolymer method combined with foam creation by H_2O_2. Some of the concrete's properties such as density, heat-resistance characters, and compressive strength were compared with traditional lightweight concretes. Especially when the foamed concretes were synthesized form fly ash by Geopolymer method, the heat-resistance of product could be as high as 1000^0C.

INTRODUCTION

In the constructive field, lightweight or foamed concretes based on the binder as Portland cement with fine sand or lightweight aggregates, the blowing agents including hydrogen peroxide, aluminum powder, polystyrene, are the traditional materials. One of the problems that relates to production process is using Portland cement, which has partly been involved in CO_2 release into the atmosphere. Furthermore, some of these products have properties such as heat-resistance, acid-resistance are not good. The foamed concretes synthesized from fly ash, sodium silicate solution and H_2O_2 as a foaming agent can significantly overcome these drawbacks.

There are two methods of producing the foamed geopolymer concretes: (1) is the thermal expansion of (Na, K) - poli(sialate-multisiloxo), with ratio Si: Al >>6, and (2) the expansion of geopolymer precondensates with blowing agents. The technology has been developed by Cordi-Géopolymere with MK-750 based on K-poly (sialate-siloxo) [1, 2, 3].The foaming agent is oxygen gas resulting from the decomposition of peroxides in alkali medium (H_2O_2, organic peroxides, sodium perborate) [1, 2, 3]. There are several concentrations of H_2O_2: 10, 30, 50, 110 volumes. For foam production the process is started with 30 vol. concentration of H_2O_2. If necessary, it can be shifted to 110 vol.

The industrial foams Trolit are manufactured with different geopolymeric raw materials such as silica fume and alumina fume[123]. Table 1 gives the physical properties of geopolymer foam, in this particular case Trolit foam, after Liefk (1999) [1, 2, 3].

Table 1: Physical properties of TROLIT foams

Bulk density [kg/m^3]	200 – 800
Max. temperature of application [C^o]	1000
Max. thermal stability [C^o]	1200
Thermal conductivity [Wm^{-1}K^{-1}]	≥ 0.037 (depending on density)
Pore parameter [mm]	0.5 – 3.0
Compressive strength [N/mm^2]	0.5 – 2.0
Shrinkage (800°C) [%]	< 1.5

Thermal conductivity is a function of the density, which means that good thermal insulating values require low density material. Figure 1 displays the variation of compressive strength with the density of foamed MK based K- poly (sialate-silixo) [1, 2, 3]. Strength varies strongly with the density and low-density foams are fragile.

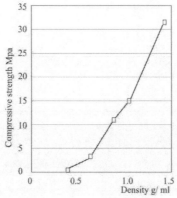

Figure 1: The Compressive strength and density of MK-750 based K – poly(sialate-siloxo)

Geopolymer foam presents a unique combination of a low thermal conductivity associated with excellent mechanical properties and very high temperature stability. These become even more obvious if the focus is put on the maximum application temperature and compared with other insulating materials (Figure 2) [1, 2, 3].

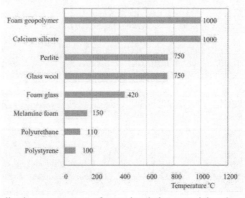

Figure 2: Maximum application temperature of some insulation materials, adapted from Liefke (1999).

EXPERIMENTAL PROCEDURES
In this study, the foamed concrete is a foamed geopolymeric concrete, which has been synthetized from some main materials such as fly ash, fine sand, sodium silicate solution, a blowing agent, and a minor of the calcium hydroxide. The blowing agent is H_2O_2 with 50 volume.

The main chemical components of the fly ash are displayed in Table 2, and the particle sizes < 45 μm make up more than 91%. The particle size distribution of the fine sand is given in table 3

Table 2: Chemical composition of fly ash (mass%)

SiO₂	Al₂O₃	Fe₂O₃	CaO	Na₂O	K₂O	TiO₂	MgO	P₂O₅	SO₃	H₂O	LOI*
43.39	29.92	10.28	2.59	0.46	0.55	1.32	1.19	1.34	0.12	-	2.97

* Loss on ignition

Table 3: Grading combination of fine sand

	Diameter (mm)						
	≥ 2	2 ÷ 1	1 ÷ 0.5	0.5÷0.25	0.25÷0.125	0.125÷0.074	≤ 0.074
Mass	0.15	6.13	27.59	28.71	23.39	6.43	8.66
Dis.in per.	0.14	6.069	27.316	28.425	23.158	6.366	8.574
Combination	100	99.86	93.82	66.51	38.09	14.94	8.574

The alkaline solution is a combination between the sodium silicate solution and the sodium hydroxide solution so that it's MR ($SiO_2 : Na_2O$) = 1.2. The alkaline solution is prepared one day prior to use.

In principle, the bubbles in geopolymer concrete paste could be formed by the addition of hydrogen peroxide. The decomposition of H_2O_2 in a high pH environment creates H_2O and atomic oxygen. It could react with organic materials in fly ash to release CO_2 gas. This is cause for geopolymer paste to expand.

The geopolymer paste has two parts: solid material and liquid solution. The solid part is a mixture of fly ash, fine sand with ratio 50:50 by mass, and a minor calcium hydroxide with 2% by mass; another part is a mixture of alkaline solution, hydrogen peroxide, and addition water. The ratio by mass between solid and liquid parts are always constant and equal 4.27. H_2O_2 solutions are mixed into the pastes with a various amount from low to high while other components of the paste are constant except addition water. The proportion of components in geopolymer concrete pastes is showed in table 4.

The mixture, after mixed together for 5 minutes, pours into the moulds of steel with dimension 15 x 15 x 15 cm, it starts to expand. The process of expansion will finish within 20 minutes. The samples are cured at ambient temperature for 4 to 8 hours, then using a bowstring by steel cut-off the odd parts of the concrete block.

After casting, all specimens are maintained at room temperature for three days. It was found that postponing the curing for periods of time causes an increase in the compressive strength of concrete [1, 2, 3]. At the end of three days, the specimens are placed inside the steam-curing chamber and cured at 60°C for 24 hours.

Table 4: Mixture proportions of foamed geopolymer concretes

Samples	Fly ash (g)	Sand (g)	Ca(OH)$_2$(g)	Alkaline Sol.(ml)	H$_2$O$_2$ (%$\sum$)	Solid/liquid (mass)
B107	307.69	307.69	12.56	100	0.09	4.27
B106	307.69	307.69	12.56	100	0.13	4.27
B105	307.69	307.69	12.56	100	0.17	4.27
B104	307.69	307.69	12.56	100	0.22	4.27
B103	307.69	307.69	12.56	100	0.26	4.27
B102	307.69	307.69	12.56	100	0.30	4.27
B101	307.69	307.69	12.56	100	0.35	4.27
B100	307.69	307.69	12.56	100	0.39	4.27
B99	307.69	307.69	12.56	100	0.43	4.27
B98	307.69	307.69	12.56	100	0.48	4.27
B97	307.69	307.69	12.56	100	0.52	4.27
B96	307.69	307.69	12.56	100	0.55	4.27
B95	307.69	307.69	12.56	100	0.61	4.27
B94	307.69	307.69	12.56	100	0.65	4.27
B89	307.69	307.69	12.56	100	0.67	4.27
B90	307.69	307.69	12.56	100	0.71	4.27
B91	307.69	307.69	12.56	100	0.74	4.27
B92	307.69	307.69	12.56	100	0.77	4.27
B93	307.69	307.69	12.56	100	0.81	4.27
B86	307.69	307.69	12.56	100	0.89	4.27
B85	307.69	307.69	12.56	100	0.00	4.27

To avoid condensation over the concrete, the concrete surface is covered by a sheet of plastic.After curing, the blocks are removed from the chamber and left to air-dry at room temperature for another 24 hours. The specimens set in the laboratory ambient conditions until the day of testing. The laboratory temperature varied between 27°C and 31°C during that period. After 28 days age, the specimens were tested the essential properties such as compressive strength, bulk density, pore parameter, maximum temperature of application, shrinkage.

RESULTS AND DISCUSSION

The interrelations between the percentage of hydrogen peroxide, the compressive strength, the bulk density and the pore parameter of these specimens are displayed in Table 5. In some cases showed that the ratio between the solid and the liquid parts of the pastes really affected to expansion and compressive strength of the geopolymer pastes.

Table 5: The interrelation between the physical properties and mass of H_2O_2

Name of Samples	H_2O_2 % $\sum$	Density (g/ml)	Compressive strength (Nmm^2)	Pore parameter (mm)	Digital pictures
B107	0.09	1.407	15.25	< 0.2	
B106	0.13	1.303	10.12	< 0.5	
B105	0.17	1.197	7.89	< 1.0	
B104	0.22	1.110	7.45	< 1.2	
B103	0.26	1.013	7.01	< 1.3	
B102	0.30	0.950	5.89	< 1.5	
B101	0.35	0.872	5.34	< 2.0	
B100	0.39	0.847	5.01	< 2.3	
B99	0.43	0.802	4.91	<2.5	
B98	0.48	0.744	4.51	<3.0	
B97	0.52	0.695	4.23	< 3.5	

B96	0.55	0.668	3.89	< 4.0	
B95	0.61	0.619	3.76	< 5.0	
B94	0.65	0.618	3.65	< 5.2	
B89	0.67	0.561	3.42	< 5.2	
B90	0.71	0.549	3.38	< 5.3	
B91	0.74	0.542	3.35	<5.5	
B92	0.77	0.518	3.00	< 6.0	
B93	0.81	Failed	Failed	Failed	
B86	0.89	failed	Failed	Failed	
B85	0.00	1.8	30	Nano size	

The test results showed that an amount of hydrogen peroxide in the geopolymer pastes is directly proportional to the expanded volume and inversely proportional to the density and compressive strength of the foamed concrete blocks. The useful limit of the H_2O_2 changes from 0.09 wt% to 0.77 wt% when mixed into the concrete pastes, and the volume of the pastes increases from 21% to 71%.

The foaming agent influences strongly on the compressive strength. It could be reduced from 90% to 50%. Therefore in the reinforced applications are disadvantage. The relations between the H_2O_2 amount, the bulk density, and the compressive strength of the foamed concretes are displayed in Figure 4 and Figure 5.

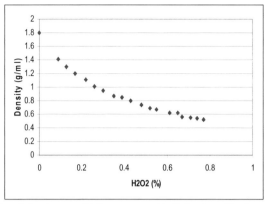

Figure 3: The relation between H_2O_2 amount and bulk density of the foamed concrete.

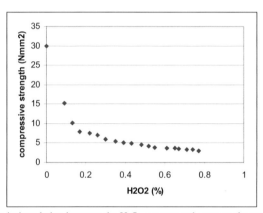

Figure 4: the relation between the H_2O_2 amount and compressive strength.

Table 6: Some of the physical properties of the foamed concrete made from fly ash

Bulk density [kgm^3]	518 - 1407
Max. temperature of application [C^o]	1000
Pore parameter [mm]	0.2 – 6.0
Compressive strength [Nmm^2]	3.0 – 15.25
Shrinkage (800^oC) [%]	< 1.0

The other physical properties of the foamed concrete were tested and displayed in table 6. The table 6 shows that the heat-resistance of the foamed concretes is higher than the traditional foamed concretes. However, in this study, the density of the foamed concretes could not reach to lower than 400 kg/m^3.

CONCLUSIONS

The article presents the first results on making foamed concrete from fly ash based on Geopolymer method combined with blowing agent hydrogen peroxide. The useful limit of the H_2O_2 50 volume is used from 0.77 wt% to 0.09 wt%. The H_2O_2 amount in the geopolymer paste is directly proportion to the expanded volume, inversely proportional to the density and the compressive strength of the foamed blocks. The capacity of foamed concrete pastes increased from 21% to 71%. The compressive strength could be reduced from 50 to 90 percent. The heat-resistance of the foamed concretes is higher than the traditional foamed concretes. It could reach to 1000 degree.

REFERENCES

[1] J. Davidovits, Geopolymer Chemistry & Applications, *Institute Geopolymer*, 469-476 (2008).

[2] E. Leifke, Industrial applications of foamed inorganic polymers, *Geopolymer '99 Proceedings*, 189 (1999).

[3] G. Schmidt, P. Randel, H-W. Engels and B. Geick, (1993), Furnace with in situ foamed insulation and process for its manufacture, PCT publication WO93/04010, US Patent 5,485,986 (1996).

[4] S. E. Wallah, B. V. Rangan, Low-calcium fly ash-based geopolymer concrete: long-term properties, Research report GC2 faculty of Engineering, *Curtin University of Technology, Perth, Australia,* 12-20 (2006).

[5] M. D. J. Sumajouw, B. V. Rangan, Low-calcium fly ash-based geopolymer concrete: reinforced beams and columns, Research report GC3 faculty of Engineering, *Curtin University of Technology, Perth, Australia*, 15-26 (2006).

PREPARATION OF ELECTRICALLY CONDUCTIVE MATERIALS BASED ON GEOPOLYMERS WITH GRAPHITE

Z. Černý, I. Jakubec, P. Bezdička, L. Šulc, J. Macháček, J. Bludská, and P. Roubíček*
Institute of Inorganic Chemistry of the ASCR, v.v.i., 250 68 Řež.
*České lupkové závody a.s., Nové Strašecí č. p. 1171, 271 11 Nové Strašecí.
Czech Republic.

ABSTRACT

Series of metakaolinite-based geopolymer composites containing from 9.1 to 44.5 wt% of graphite were prepared by mixing of metakaolinite, sodium silicate solution (water glass) and graphite powder. Prepared materials were investigated by XRD, SEM, and DTA measurements.

The increasing content of graphite in the composite increases the electric conductivity of the material, measured by Four Point Resistivity method. The specific conductivity of the prepared materials is stable up to 300°C and almost independent on temperature. The resulting materials can be copperized to get a „ cupriferous like body".

INTRODUCTION

Geopolymers represent mainly inorganic amorphous crosslinked polymers, which are usually formed by reaction between a solid aluminosilicate (metakaoline) and sodium silicate solution (water glass).[1-3] Main effort of using of the geopolymer matrix in larger scale is focused on the competition of geopolymer materials with materials based on Portland cements.[4] Quite a different approach is represented by using of the geopolymers as highly qualified matrixes. Such matrixes can be filled with TiO_2 pigment for preparation of photocatalytic layers, by graphite, polytetrafluorethylene and molybdenum disulfide for preparation of materials with increased tribological properties, by carbon or glass fibers or by polyethylene or polypropylene powders.[5-7]

A pure geopolymer matrix prepared at room temperature exhibits mainly ionic electrical conductivity, however, the material becomes a typical insulator after thermal curing of the materials over 300°C.[8] The present work describes preparation of new thermally stable electrically conductive materials.

EXPERIMENTAL

Geopolymer materials are generally based on alkalic activation of dehydroxylated kaolinite - metakaolinite. In this work the geopolymer matrix was prepared directly from silicate anions (sodium silicate solution) and metakaolinite mixed "in situ" with graphite powder at room temperature.

Chemicals and raw materials

Sodium silicate solution - water glass, supplied by Koma Ltd. Praha, Czech Republic, analytical composition: 27 wt% SiO_2, 9.1 wt% Na_2O, density = 1.38 g/cm3, content of Fe_2O_3 < 1 wt%.

Metakaolinite, MEFISTO KO5 product, České lupkové závody, Corp., Nové Strašecí, Czech Republic, www.cluz.cz., analytical composition: 40.7 wt% Al_2O_3, 57.7 wt% SiO_2, 0.53 wt% TiO_2 and 0.58 wt% Fe_2O_3.

Graphite powder supplied by KOH-I-NOOR GRAFIT Ltd., www.grafitnetolice.cz, commercially available CR5 995, content of C = 99.5 wt%, content of ash < 0.5 wt%, humidity < 0.4 wt%, d_{50} < 5.5 - 7 μm, specific surface =10 m^2/g.

All these above mentioned products were used without any other treatment.

Characterization methods

Diffraction patterns were collected by the PANalytical X´Pert PRO diffractometer equipped with conventional X-ray tube (Cu K radiation, 40 kV, 30 mA) and a linear position sensitive detector PIXcel with an anti-scatter shield. A programmable divergence slit set to a fixed value of 0.5 deg, Soller slits of 0.02 rad. and mask of 15 mm were used in the primary beam. A programmable anti-scatter slit set to fixed value of 0.5 deg., Soller slit of 0.02 rad and Ni beta-filter were used in the diffracted beam. Data were taken in the range of 5 - 100 deg 2theta with the step of 0.0131 deg and 200s / step. Qualitative analysis was performed with HighScorePlus software package (PANalytical, the Netherlands, version 2.2.5), Diffrac-Plus software package (Bruker AXS, Germany, version 8.0) and JCPDS PDF-2 database [1].

SEM (scanning electron microscopy) studies were obtained by Philips XL30 CP microscope at 30 kV.

DTA and TG were obtained by NETZSCH apparatus STA 409 for a simultaneous TG and DTA analysis. The measurements were carried out between cca. 20 and 1400°C in inert gas atmosphere with intelligent coupling systems for QMS for evolved gas analysis during the process. Rate of heating was equal to 10 °C/min.

Specific electric conductivity values of the combined materials were obtained using the four-point probe station consisting of four probe tips, an amperemeter (Metra), DC current source (potentiostat/galvanostat PS4 Forschunginstitut Mainsberg), and a voltmeter (KEITHLY2000 multimeter). The four probes were arranged in a linear way, in which two outer probes were connected to a current supply, and two inner probes to a voltmeter. As current flows between the outer probes, the dropped voltage on the inner probes was measured. The relationship between the values of current and voltage depends on the resistivity of the examined material and the geometrical characteristics of the probe as per follows:

$$= I/U * 1/(h*F(h/s)*F´(d/s))$$

where stands for the specific conductivity of the material, I for the current, U for the potential. F(h/s) and F´(d/s) represent geometrical factors of relationships between the height (h) and diameter of the sample (d) and linear distance of the probe tips (s). The samples for electrical characterization were kept for 30 min at given temperature before the measurements of specific conductivity.

Water extracted portions of the composite materials were determined at 22°C by immersing of bodies into water for 24 hours; after that the bodies were dried at ambient temperature to constant mass.

Equilibrium sorption of water of the material was determined on water extracted bodies by immersing of the bodies into the water to constant mass at 22°C.

Densities of the bodies were determined by hydrostatic method on bodies saturated by water at ambient temperature.

Values of water extracted portions, equilibrium absorptions and the densities represent an average of three independent measurements.

Preparation procedure

A pregeopolymer suspension was prepared by mixing of 240g of water glass and 80g of metakaolinite in well stirred 1 litre vessel for 10 minutes at ambient temperature.

Composite graphite materials were prepared by mixing of a given amount of graphite powder directly into the above pregeopolymer suspension and mixed for another 10 min. The blended mixtures were placed into disk shape moulds with diameter of 25 and 1.8 mm prepared from silicone rubber. The samples were allowed at ambient temperature to constant mass. In this way, bodies of samples of the combined materials with a typical size of 23 - 24 mm in diameter and thickness of 1.4 - 1,5 mm of were obtained.

A series of samples containing 9.1, 16.8, 28.6, 37.5 and 44.5 wt% of graphite, calculated for constant mass of bodies at 300°C, were prepared by the above procedure.

Combined materials containing 16.8% of graphite were subjected to an electrochemical copperizing done in a bath prepared from 200g $CuSO_4$. $5H_2O$, 0.5g H_2SO_4 and 0.4g thiocarbamide per 1 liter of the mixture for 2 hours at 1- 4 A/dm^2 with content.

RESULTS AND DISCUSSION

a) Characterization of the initial graphite powder is introduced in Figure 1-3. According to SEM, Figure 1., the initial graphite shows a flaky structure in agreement with certificated parameters of $d_{10} \sim 2.5$ μm and $d_{50} \sim 6$ μm for commercial type CR5. Results of DTA , Figure 2., corresponds to certificated purity > 99.6 %. Oxidation of initial graphite starts at about 570°C and graphite completely burn up at about 1320°C under given conditions. XRD record of the graphite powder, Figure 3., shows a typical diffraction pattern.

Figure 1. SEM of the initial raw graphite powder CR 5

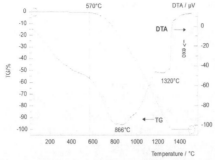

Figure 2. DTA of the initial raw graphite powder CR 5

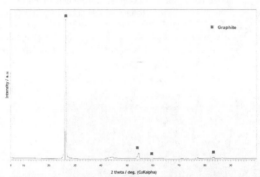

Figure 3. XRD characteristic pattern of the initial raw graphite powder CR 5

b) Morphology and XRD characterizations of the neat initial geopolymer matrix was published earlier.[5]

c) Basic characterization of the series of the graphite composites is shown in Figures 4. - 6. An increase in sorption of the water in the composites from 3% (for material with 9.1% of graphite) to more than 25% (for material with 44.5% of graphite), Figure 4., indicates an increasing portion of opened porosity (available for water) with increasing content of graphite in the materials, (see internal porous character of the material with 28.6 % of graphite, Figure 10.). On the other hand, density, Figure 5., hydrostatically determined on water-saturated bodies of the materials, decreases with increasing content of graphite in spite of the fact that bulk density of graphite is about 2.25 g/cm^3. This finding suggests that some part of the internal porosity in the bodies remains closed to the water sorption. Water extracted portions, Figure 6., decrease with decreasing content of geopolymer matrix to minimum for 28.6 % of graphite; the following increase of the extracted portions could be explained by scrubbing of some portion of the graphite particles by the water.

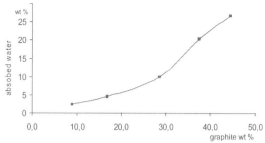

Figure 4. Dependence of equilibrium sorption of water of the combined materials on content of graphite at 22°C.

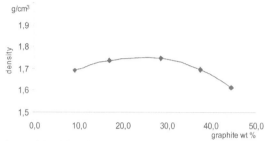

Figure 5. Dependence of densities of the combined materials (saturated by water) on content of graphite at 22°C

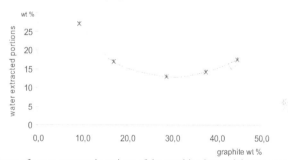

Figure 6. Dependence of water extracted portions of the combined materials on content of graphite at 22°C.

Composite materials with 9.1% and 44.5% of graphite are characterized in Figures 7. – 9. SEM micrographs on Figure 7. show expected rough structure of the materials obviously containing flaky relicts of the graphite. DTA records, Figure 8., indicate that both materials contain 11 – 13 % of bonded water released above 100°C, see the endothermic peaks at 104°C (A) and at 120°C (B), respectively. The temperature corresponding with the upset of oxidation of the graphite decreases significantly from 494°C for 9,1% to 385°C for 44.5% of graphite in a composites; compare this values with the 570°C for raw graphite, Figure 2. The changes correspond to an increase of the opened porosity of the material with increasing content of graphite. Maximums of exothermic peaks, at 795°C (A) and 790°C (B), respectively, correspond to burning-out of graphite in both materials. Mass losses of the bodies after combustion of graphite were in full accordance with the calculated content of graphite in material dried at 350°C. XRD records, Figure 9., of the material with 9.1% of graphite shows diffraction patterns consistent with main features of initial metakaoline, containing a quartz impurity and graphite superimposed in one main peak and a trace of corundum. The same features contain XRD patterns of material with 44.5% of graphite. From semi quantitative analysis of the XRD patterns can be estimated amorphous portion ≥ 95% for both materials.

Figure 7. SEM characterization of combined materials with 9.1 % (A) and 44.5 % (B) of graphite

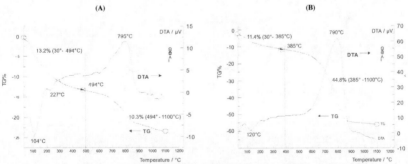

Figure 8. DTA characterization of combined materials with 9.1 % (A) and 44.5 % (B) of graphite

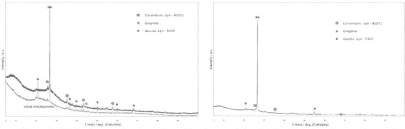

Figure 9. XRD characterization of combined materials with 9.1 % (A) and 44.5 % (B) of graphite.

While insides of the materials represent highly opened porous structures, Figure 10.(A), on the outer surfaces of the bodies with higher content of graphite an efflorescence was found after a few weeks, typically after 3 weeks at ambient temperature, Figure 10.(B). The efflorescence comes from reaction of residual NaOH diffused on surface of the material with atmospheric CO_2 resulting into hydrated sodium salts of carbonate, that were identified by XRD as thermonatrite, Na_2CO_3, and hydrocarbonate, trona, $Na_3H(CO_3)_2$, Figure 11.

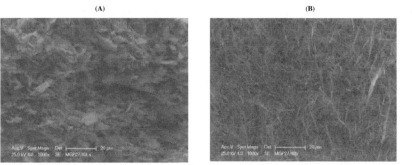

Figure 10. SEM micrographs of the internal macro-porous character (A) and efflorescence on the surface (B) of the material with 28.6 % of graphite.

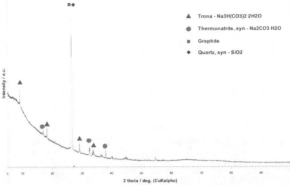

Figure 11. XRD pattern of the efflorescence on the surface of the material with 28.6 % of graphite.

d) While neat geopolymer matrix dried at 300 - 350°C is an insulator, the geopolymers modified by graphite represent inherently electrically conducting materials characterized by specific electric conductivity, Figure 12. The values of the specific conductivity increase with the content of graphite and are almost independent on temperature or slightly increase due to increasing conductivity of the graphite with increasing temperature. The material is stable up to 300°C, above this temperature the graphite starts slowly burn up and material is loosing the conductivity.

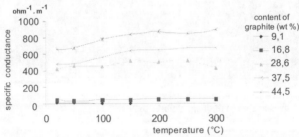

Figure 12. Dependence of specific conductivity of the composite graphite materials on temperature.

e) The well conducting composite graphite materials can be treated by electrochemical copperizing, resulting in „ cupriferous like bodies" covered by thin rough layer of Cu reflecting the initial uneven surface of the geopolymer matrix, Figure 13.

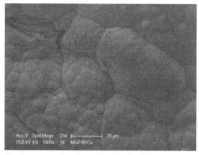

Figure 13. SEM macrographs of chemically copperized composite bodies

CONCLUSION

In present work the characterization of material based on geopolymer matrix and graphite has been studied. The new composite material exhibits internal electric conductivity, the specific conductivity of which increases with the content of graphite. The conductivity is very stable and almost independent on temperature in range 22 - 300°C. Over 350°C an oxidation of the graphite occures in the materials; the temperature of the of the oxidation dereases from 570°C for raw graphite to temperature under 400°C for materials with higher content of graphite.

The material can be copperized to get a „cuprifferous like body", however the Cu surface reflects rough surface of the initial geopolymer matrix.

ACKNOWLEDGMENTS

The authors thank the Academy of Science of the Czech Republic for project AV0Z40320502 and Ministry of Industry and Trade of the Czech Republic for project 2A-1TP1/063.

REFERENCES

[1]J. Davidovits, and J. Davidovits, Geopolymers: inorganic polymeric new materials, *J. Therm. Anal.* **37**, 1633–1656 (1991).

[2]H. Xu, J. S. J. Van Deventer, The geopolymerisation of alumino-silicate minerals, *Int. J. Miner. Process.* **59**, 247-266 (2000).

[3]A. Buchwald, H. Hilbig, and C. Kaps, Alkali-activated metakaolin-slag blends - performance and structure in dependence of their composition, *J. Mater. Sci.*, **42**, 3024–3032 (2007).

[4]J. Davidovits, Geopolymer chemistry and application, published by Institut Geopolymere, Saint-Quentin, France, (2009).

[5]Z. Cerny, I. Jakubec, P. Bezdicka, et al.: Preparation of photocatalytic layers based on geopolymer, *Ceramic Engineering and Science Proceedings*, **29**, 113-121 (2009).

[6]H-L. Wang, H-H. Li, F-Y. Yan, Reduction in wear of metakaolinite-based geopolymer composite through filling of PTFE, *Wear*, **258**, 1562-1566 (2005).

[7]H-L. Wang, H-H. Li, F-Y. Yan, Synthesis and tribological behavior of metakaolinite-based geopolymer composites, *Mater. Lett.* **59**, 3976-3981 (2005).

[8]X-M. Cui, G-J. Zheng, Y-C. Han, F. Su, J. Zhou, A study on electrical conductivity of chemosynthetic Al2O3-2SiO(2) geoploymer materials, *J. Pow. Sour.* **184**, 652-656 (2008).

EFFECT OF SYNTHESIS PARAMETERS AND POST-CURE TEMPERATURE ON THE MECHANICAL PROPERTIES OF GEOPOLYMERS CONTAINING SLAG

Tammy L. Metroke[1], Brian Evans[1], Jeff Eichler[1], Michael I. Hammons[2], Michael V. Henley[2]
[1]Universal Technology Corporation, 139 Barnes, Suite 2, Tyndall AFB, FL 32403
[2]Air Force Research Laboratory, Airbase Technologies Division, 139 Barnes, Suite 2, Tyndall AFB, FL 32403

ABSTRACT

The effect of ground granulated blast furnace slag (GGBFS) preparation method (wet, dry, milled) on the mechanical strength of fly ash (FA)-based geopolymers was determined as a function of post-cure thermal exposure temperature. Slag preparation method affected the mechanical strength of FA–GGBFS geopolymers containing 20–40 % GGBFS, wet > dry > milled. FA-based geopolymers containing 20–40 % wet or dry GGBFS or 20 % milled GGBFS exhibited higher compressive strengths than the binder materials at temperatures $\leq 1000°C$. Results indicate that the addition of GGBFS led to a decrease in the extent of thermal stress cracking of FA-based geopolymers after thermal exposure to $1000°C$. Slag preparation method likely influences the rate of release of calcium species from the slag and may affect the calcium silicate–calcium aluminosilicate gel ratio in the geopolymer product, the amount of slag available to function as a microaggregate, the porosity characteristics of the gel product, and the material's mechanical strength.

INTRODUCTION

Although metallurgical slags are not as well-characterized as metakaolin (MK) or fly ash (FA), recent studies have shown that they may be used to produce geopolymers. Ground granulated blast furnace slag (GGBFS), or slag, is a calcium aluminosilicate[1] comprising glassy particles of regular chemical homogeneity but irregular morphology. Slag contains a significant fraction of vitreous material that is highly reactive towards geopolymerization, especially when used in combination with other aluminosilicate source materials. For example, geopolymers were prepared from mixtures of kaolinite–low calcium ferronickel slag[2], MK–slag[3], and Class C FA–blast furnace slag.[4] The abrasion[5], fire[6], and erosion[4] resistance of geopolymers containing blast furnace slag were determined. Yunsheng et al. investigated the synthesis and heavy metal immobilization behaviors of a geopolymer prepared from MK–slag mixtures as a function of slag content and curing method[3]. Yip et al. investigated the coexistence of geopolymeric gel and calcium silicate hydrate (CSH) in MK–GGBFS geopolymers.[7]

This paper presents a systematic determination of the effect of GGBFS addition on the physical and mechanical properties of geopolymers prepared from FA. The effects of slag content, slag preparation method (wet, dry, milled) and post-cure thermal exposure on the mechanical strength and thermal stability of geopolymers were determined.

EXPERIMENTAL

Materials

The aluminosilicate source material was class F fly ash (FA) (Boral Material Technologies). FA contained 53.94 % SiO_2, 28.25 % Al_2O_3, and 7.29 % Fe_2O_3. GGBFS (Holcim Ltd.) contained 37.75 % SiO_2, 8.73 % Al_2O_3, 0.37 % Fe_2O_3, 36.42 % CaO, 13.69 % MgO, and 0.43 % Mn_2O_3. Potassium silicate Kasil 6 (PQ Corporation) was used as the silicate. Kasil 6 contains 26.5 % SiO_2 and 12.65 % K_2O. Reagent grade potassium hydroxide (Aldrich) was used.

Slag Preparation

Slag was incorporated into the geopolymer in one of three states: wet, dry, or milled. Wet slag contained approximately 10–12 % water and was used as received from the manufacturer. Dry slag was prepared by heating wet slag in a 180°C oven overnight. Milled slag was prepared by grinding dry slag in a ball mill for 2 days. Table 1 describes the particle size distribution for the GGBFS. For the purposes of this study, it is assumed that the particle size distributions for wet and dry slag were similar.

Geopolymer Synthesis

FA geopolymers were prepared using $SiO_2/Al_2O_3 = 4$ and $H_2O/M_2O = 2$. The K_2O/SiO_2 of the silicate solutions used to prepare the geopolymer materials was 1.72. In a typical

Table 1: Particle size distribution of dry and milled GGBFS. The numbers represent the percentage exceeding the sieve size.

Sieve Size (μm)	Dry GGBFS	Milled GGBFS
2000	0	0
1400	10.4	0
600	28.1	2.2
300	72.7	4.0
250	91.7	34.8
150	92.6	40.3
75	96.6	58.4

preparation, the potassium silicate and potassium hydroxide were mixed for a 24-hr period prior to use. FA was then added with vigorous mixing using a rotary mixer for 10 minutes to form the geopolymer binder. Slag was incorporated by hand mixing, except in the case of the dry milled slag. Dry milled slag was blended into the geopolymer binder using the rotary mixer for 2–3 minutes. The materials were transferred into plastic molds, covered with a plastic top, allowed to sit at room temperature overnight, and cured at 90°C for 5 days. After curing, the materials were cooled and held under ambient conditions prior to analysis. The amount of slag incorporated ranged from 20–80 wt. % and varied by slag preparation method.

Post-Cure Thermal Exposure

After curing at 90°C, the materials were placed in an oven at 300, 500, 800, or 1000°C for 1 hour. After post-cure thermal exposure, the materials were allowed to cool in air to room temperature prior to analysis.

Mechanical Strength Analysis

The compressive strengths of the cured geopolymers were determined using an ELE compression tester operated using a 200 lbf/s strain rate in accordance with ASTM C109. Test specimens were 2 x 2 x 2 inch cubes. All values presented in the current work are an average of results obtained for three samples.

RESULTS AND DISCUSSION

Figure 1 shows the physical appearance of fly ash-based geopolymers prepared using wet, dry, or milled GGBFS after curing at 90°C and post-cure thermal exposure to 500°C or 1000°C. Wet slag was readily incorporated in concentrations of 20–80 % into the fly ash-based binder. Good quality materials were prepared using 20–60 % dry slag and 20–40 % milled slag. The binder material exhibited signs of cracking after heating to 1000°C, while materials containing GGBFS exhibited no appreciable cracking of the material surface after thermal exposure under the same conditions.

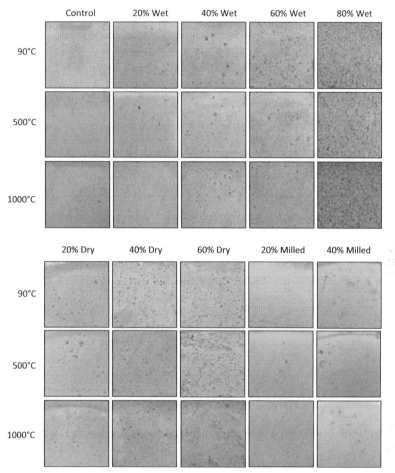

Figure 1: Photos of geopolymers containing various concentrations of wet, dry, or milled slag after curing at 90°C or heating to 500°C or 1000°C.

Figure 2 shows compressive strength as a function of GGBFS content and preparation method for the FA-based materials. In general, the addition of 20–40 % wet or dry slag had a marginal affect on material strength, with the strengths ranging from 6160–7950 psi. The addition of 60–80 % wet or dry slag led to a decrease in strength to approximately 980–3270 psi. For FA–GGBFS materials, the addition of milled slag reduced compressive strength to 5050 and 4000 psi for materials containing 20 and 40 % milled slag, respectively.

Figure 3 shows residual compressive strength as a function of post-cure thermal exposure for FA-based geopolymers containing wet, dry, or milled GGBFS. For the binder and the materials containing wet slag, compressive strength was found to decrease rapidly with increasing temperature, though the strength of the FA–GGBFS materials containing 20–40 % GGBFS was comparable to or higher than the binder after heating to 300–500°C. The strength of materials containing 60 % wet slag remained relatively constant at approximately 3100 psi until the post-cure temperature reached 500°C, then decreased. The strength of the material containing 80 % wet GGBFS was low, approximately 1000 psi, and remained relatively constant during thermal exposure.

The residual strength of the materials containing 20–40 % dry GGBFS was greater than the binder and remained relatively constant upon heating to 300°C, but decreased upon further heating to 500°C. After heating to 1000°C, the strength of the materials containing 20 or 40 % dry GGBFS was 1600 and 3100 psi, respectively. The strength of the material containing 60 % dry GGBFS remained relatively constant at approximately 1700 psi during thermal exposure to 800°C, and increased to 2130 psi upon heating to 1000°C.

For the FA–GGBFS material containing 20 % milled slag, the residual strength after heating at 300°C was

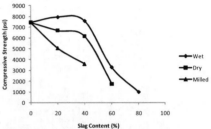

Figure 2: Compressive strength as a function of slag content and slag preparation method for FA-based geopolymers containing GGBFS.

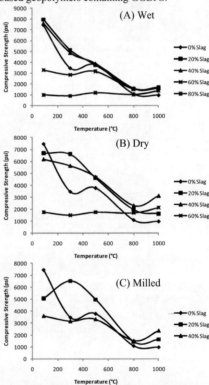

Figure 3: Residual compressive strength as a function of temperature for FA-GGBFS materials prepared from (A) wet, (B) dry, or (C) milled GGBFS.

approximately 6500 psi and decreased to 4960 psi upon heating to 500°C, but both were stronger than the binder material at equivalent temperatures. The strength of the FA-based material containing 40 % milled slag was lower, approximately 3200 psi, and remained relatively constant for temperature $\leq$ 500°C.

Previous studies have shown that the addition of GGBFS to metakaolin[8]- or FA-based[9] geopolymers results in the formation of calcium silicate (CS) or calcium aluminosilicate (CAS) gel, depending on the release of AlO_4^- tetrahedra from the aluminosilicate source material. In the FA–GGBFS system, we expect the release of AlO_4^- relative to the release of calcium species from the slag to be low, likely resulting in the formation of a calcium silicate gel product. The slag preparation method likely influences the rate of release of calcium species from the slag and may affect the CS–CAS ratio in the geopolymer gel product, the amount of slag available to function as a microaggregate, the porosity characteristics of the gel product, and the material's mechanical strength. As observed by Kong and Sanjayan[10] and Pan et al.[11], thermal exposure resulted in a decrease in residual compressive strength for both the binder and the slag-containing materials, likely due to dehydration/dehydroxylation, phase changes, and porosity variations caused by the elevated temperature.

CONCLUSIONS

Slag preparation method affected the mechanical strength of FA–GGBFS geopolymers containing 20–40 % GGBFS, wet > dry > milled. Slag incorporation reduced the extent and size of cracks due to thermal stresses observed after post-cure thermal exposure to 1000°C. For FA-based materials, incorporation of 20–40 % (wet, dry) or 20 % milled GGBFS resulted in materials exhibiting higher compressive strength after post-cure thermal exposure to 300–800°C than the analogous binders. From these results, we conclude that the addition of GGBFS to geopolymers increases the material's thermal stability; the overall material strength is likely related to slag preparation method, slag content, post-cure thermal exposure temperature, and microstructural properties of the resulting geopolymer gel product.

ACKNOWLEDGEMENTS

The authors would like to thank Professor Waltrud "Trudy" Kriven for helpful discussions.

REFERENCES
[1] N. Tsuyuki, K. Koizumi, Granularity and Surface Structure of Ground Granulated Blast-Furnace Slags, *J. Am. Ceram. Soc.*, **82**, 2188–2192 (1999).

[2] K. Komnitsas, D. Zaharaki, V. Perdikatsis, Geopolymerisation of Low Calcium Ferronickel Slags, *J. Mater. Sci.*, **42**, 3073–3082 (2007).

[3] Z. Yunsheng, S. Wei, C. Qianli, C. Lin, Synthesis and Immobilization Behaviors of Slag Based Geopolymer, *J. Hazard. Mat.*, **143**, 206–213 (2007).

[4] K.C. Goretta, N. Chen, F. Gutierrez-Mora, J.L. Routbort, G.C. Lukey, J.S.J. van Deventer, Solid-Particle Erosion of a Geopolymer Containing Fly Ash and Blast-Furnace Slag, *Wear*, **256**, 714–719 (2004).

[5] S. Hu, J. Wang, G. Zhang, Q. Ding, Bonding and Abrasion Resistance of Geopolymeric Repair Material Made with Steel Slag, *Cement & Concrete Composites*, **30**, 239–244 (2008).

[6] T.W. Cheng, J.P. Chiu, Fire-Resistant Geopolymer Produced by Granulated Blast Furnace Slag, *Miner. Eng.*, **16**, 205-210 (2003).

[7] C.K. Yip, G.C. Lukey, J.S.J. van Deventer, The Coexistence of Geopolymeric Gel and Calcium Silicate Hydrate at the Early Stage of Alkaline Activation, *Cement Concrete Res.*, **35**, 1688–1697 (2005).

Computational Design, Modeling, Simulation and Characterization

ELECTRONIC STRUCTURE AND BAND-GAPS OF Eu-DOPED LaSi₃N₅ TERNARY NITRIDES

L. Benco[1,2,] Z. Lences[1], P. Sajgalik[1]
[1]Institute of Inorganic Chemistry, Slovak Academy of Sciences, Dubravska cesta 9, 84536 Bratislava, Slovakia
[2]Computational Materials Physics and Center for Computational Materials Science, Sensengasse 8, 1090 Vienna, Austria

ABSTRACT

The Eu-doped $LaSi_3N_5$ has luminescence in the blue-green light region. With increasing europium content in general formula $La_{1-z}Eu_zSi_3N_{5-z}O_{3/2z}$ from $z = 0.05$ to $z = 0.1$ the emission intensity increases, indicating the increase of the concentration of the luminescent centres. First-principles Density-Functional Theory (DFT) calculations are performed to enhance the understanding of the electronic structure of the stoichiometric $LaSi_3N_5$ and La/Eu and N/O substituted ternary nitrides. To mimic the realistic concentration of Eu and O the cell volume is expanded to the 2x1x2 super-cell with 144 atoms. The calculations show that ternary nitrides are large-gap insulators. The La^{3+}/Eu^{3+} substitution does not lead to a significant change of the gap. The observed concentration of O atoms in the framework can be compensated either by the creation of the M^{3+} hole (M=La, Eu), or by the La^{3+}/Eu^{2+} substitution. Both phenomena lead to a decrease of the gap. Upon the hole creation the DFT-calculated gap decreases from ~3.15 eV to ~2.55 eV. The La^{3+}/Eu^{2+} substitution leads to the location of the narrow band of Eu 4f-states inside the gap thus narrowing the band-gap to 0.542 eV. This value of the DFT band gap is typically too small. A tuning of the position of the band of the nonbonding 4f-states away from the conduction band towards more negative energies using the LDA+U approach leads to the enlargement of the band gap up to the value of ~3 eV. Finally, a benchmark calculation using a hybrid functional provides the band-gap of 3.1 eV in reasonable agreement with experimental data. Our calculations show that in $La/EuSi_3N_5$ ternary nitrides a narrowing of the band gap is caused by the creation of the M^{3+} holes (M=La, Eu) or by the La^{3+}/Eu^{2+} substitution, both phenomena are enabled by N/O substitutions in the framework of ternary nitrides.

INTRODUCTION

The porous structure of silicon nitrides and their high thermal and chemical stability allow the formation of interstitial ternary silicon nitrides with extra-framework cations. Such compounds exhibit a good stability when interstitial sites are occupied with large and bulky cations. Inoue et al.[1] synthesised and characterized lanthanum silicon nitride, $LaSi_3N_5$, in which the interstitial La^{3+} cation compensates the excess negative charge of the silicon-nitride framework. Woike and Jeitschko[2] reported crystal structures of isotypic compounds MSi_3N_5 (M = La, Ce, Pr, Nd), and $M_3Si_6N_{11}$ (M = La, Ce, Pr, Nd, Sm). Upon doping of $LaSi_3N_5$ with Eu^{3+} a luminiscent compound is formed, which exhibits the increase of the emission intensity with increasing europium content in general formula $La_{1-z}Eu_zSi_3N_{5-z}O_{3/2z}$ from z = 0.05 to z = 0.1. In this work ab initio DFT calculations are used to investigate the electronic structure of $LaSi_3N_5$ and corresponding Eu-doped materials. A role of the Eu concentration, as well as oxidation state of Eu is studied. In order to reasonably describe the bonding of f-electrons various types of Eu pseudopotentials are used. The standard DFT is completed with a sophisticated LDA+U approach and finally, calculated band-gaps are compared with those from the hybrid-functional calculation.

STRUCTURE AND COMPUTATIONAL DETAILS

The structure of $LaSi_3N_5$ is reconstructed according to crystallographic data by Inoue et al.[1] and Hatfield et al.[3]. The lattice parameters are a = 7.833 Å, b = 11.236 Å, and c = 4.807 Å. The 9 irreducible atomic sites (1La, 3Si, 5N) multiplied by symmetry operations of the space group P $2_1 2_1 2_1$

produce 36 atomic positions. To mimic realistic concentrations of Eu and O a 2x1x2 supercell is created containing 144 atoms. The supercell is displayed in Fig. 1. The nitridosilicate framework consists of SiN$_4$ tetrahedra (not displayed) linked by sharing corners to form rings of five tetrahedral units. The lanthanum ion is located at the center of pentagonal space[1]. The Eu(III)–doping is performed via the substitution of the La^{3+} cation with the trivalent Eu^{3+} cation. In the Eu(II)–doping the Eu^{2+} cation replaces the La^{3+} cation and the closest N atom in the framework is replaced with the O atom (Fig. 2). At multiple doping the Eu cations are placed in different pentagonal spaces at large distances.

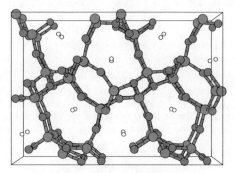

Figure 1. The 2x1x2 supercell of the LaSi$_3$N$_5$ structure (brown balls = Si, blue balls = N, empty balls = La).

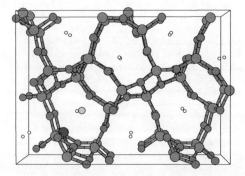

Figure 2. The cell of Eu-doped LaSi$_3$N$_5$ (yellow ball = Eu^{2+}, red ball = O, other symbols are same as in Fig. 1).

Periodic ab initio total energy calculations are performed for the relaxation of atomic positions. Our DFT-based calculations[4] use the generalized-gradient approximation[5] to the exchange-correlation functional. The standard GGA PW91 functional[5] is used to calculate the electronic structure of both the stoichiometric and Eu-doped LaSi$_3$N$_5$. A hybrid functional mixing exact and the DFT exchange[6] is applied to the bonding of the Eu^{2+} cation to obtain more realistic band-gaps of the doped material. We use ultrasoft pseudopotentials[7] and a plane-wave (PW) basis as implemented in the Vienna ab initio simulation package VASP[8-11]. The advantage of PWs is an easily tunable precision by means of the plane-wave cutoff energy, and the fact that the PW basis sets exhibit no basis set superposition error (BSSE). The calculations are performed using Blöchl's projector augmented wave technique[12,13]. Calculations use the plane-wave cutoff energy of 400 eV. Unit cells used to calculate the electronic structure of the LaSi$_3$N$_5$ phosphor are relatively large with lattice vectors of the length $\geq$ 10 Å and the cell volume larger than 1000 Å^3. The sampling of the corresponding cell in the reciprocal space (Brillouin-zone sampling) is therefore restricted to a single point (gamma-point). The convergence is improved using a modest smearing of eigenvalues. The full relaxation of atomic geometries, using a conjugate-gradient algorithm, applies a stopping criterion of 10^{-5} eV for the self-consistency loop and 10^{-4} eV for the optimizer. In the relaxation procedure no symmetry restrictions are applied. The LDA+U approach is used to introduce a strong intra-atomic interaction between f-electrons of Eu, thus shifting the position of the f-band and increasing the value of the band gap[14]. A hybrid functional suggested by Perdew and coworkers[15] is used to calculated the realistic electronic structure of Eu-doped LaSi$_3$N$_5$.

RESULTS AND DISCUSSION

Stoichiometric LaSi$_3$N$_5$

The equilibrium cell volume of the stoichiometric LaSi$_3$N$_5$ is determined using the Birch-Murnagham fit. The lattice vectors of the optimized cell and the cell volume are listed in Table 1 along with experimentally determined cell parameters and those of the Eu-doped phosphors.

Table 1. Lattice parameters and the cell volume of the LaSi$_3$N$_5$ and Eu-doped phosphors (vectors in Å, volume in Å^3)

	A	B	C	V
experimental[1]	7.838	11.236	4.807	423.342
calculated	7.897	11.326	4.848	433.552
Eu(II)-doped[*]	7.903	11.333	4.858	434.353
Eu(III)-doped[*]	7.896	11.318	4.843	433.100

[*]1Eu atom per 2x1x2 supercell.

The calculated lattice vectors of LaSi$_3$N$_5$ are slightly larger than experimentally determined values, as typical for the GGA approach. With the Eu(II)-doping the cell volume slightly expands. On the contrary, with the Eu(III)-doping the cell volume contracts. This variation goes in agreement with ionic radii of ions (cf. Table 2). The Eu(III) ion is slightly smaller than La(III) ion. The Eu(III)/La(III) substitution therefore leads to a slight contraction. In the Eu(II) doping the decrease of valency of the extra-framework cation La(III)→Eu(II) is compensated by the N/O substitution in the framework. Because Eu(II) is larger than La(III) and O(-II) is larger than N(-III), the Eu(II) doping causes the stretching of the lattice vectors and the expansion of the unit cell.

Table 2. Pauling ionic radii of La, Eu, N and O (pm)

La(III)	Eu(III)	Eu(II)	N(-III)	O(-II)
130(8)[*]	121(8)	139(8)	132	140

[*]Number in parentheses is the coordination number of the ion.

The electronic structure of LaSi$_3$N$_5$ is displayed in Figure 3. The valence energy levels are collected in the s–, p–, and the conduction band (CB). The s–band is centered at ~ –15 eV. A narrow band at ~ –32 eV is composed of deep nonbonding La 5s^2 energy levels (the pseudopotential of La is constructed for the atomic configuration 5s^2 5p^6 5d^1 6s^2 4f^0). The s–band centered at ~ –15 eV is composed mainly of s-orbitals of N, s-orbitals of Si and p-orbitals of La. The broad p–band extends between ~ –10 eV and the Fermi level. It is composed of N p-orbitals, Si s-, and p-orbitals. The bottom of the conduction band at ~3 eV consists of La f states forming a narrow band. The empty La d energy levels extend between ~3eV nad ~6eV. The large band gap of ~3.15 eV (the DFT GGA value is usually represents less than 50% of the real gap) indicates that stoichiometric LaSi$_3$N$_5$ is an insulating material.

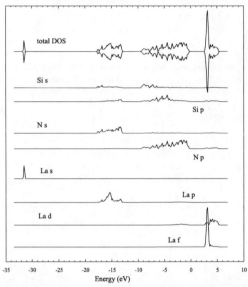

Figure 3. Total DOS of the stoichiometric LaSi$_3$N$_5$ and Si-, N- and La-projected orbital components.

With the variation of the cell volume a change of the band gap is observed. The band gap as a function of the cell volume is displayed in Figure 4. An isomorphic change of the cell volume up to ~9% to both sides of the equilibrium volume leads to a variation of the band gap between ~2.9 eV and ~3.4 eV. Interestingly, with the decrease of the cell volume (the increase of pressure) the band gap increases.

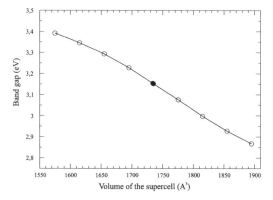

Figure 4. The band gap of LaSi$_3$N$_5$ as a function of the cell volume (the full circle indicates the gap at the equilibrium volume).

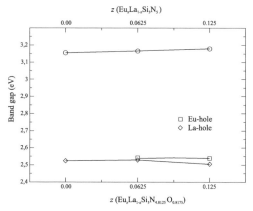

Figure 5. The band gap of the Eu(III) doped LaSi$_3$N$_5$ as a function of the La(III)/Eu(III) substitution rate (top line) and the formation of the La(III) and/or Eu(III) vacancy (bottom lines).

Eu-doped LaSi$_3$N$_5$

The effect of the substitution of the extra-framework La(III) cation with trivalent Eu(III) on the band gap is displayed in Figure 5. Upon the La(III)/Eu(III) substitution the large band gap is retained and the increasing substitution rate does not lead to a significant change of the band gap (Fig. 5, top line). Eu-doped LaSi$_3$N$_5$ phosphors always contain an admixture of oxygen atoms. In the electronic structure of O-doped materials the excess electron density introduced by O-atoms fills the energy levels of the conduction band. Such an O-doped material would exhibit conductive properties, what is not observed experimentally. A possible mechanism of the compensation of the excess electron density, introduced by O-atoms is, e.g., the creation of the M(III) cation vacancy. The effect of the creation of the M(III) vacancy is displayed in Fig. 5. For the LaSi$_3$N$_5$ phosphor with both the Eu(III)- and the

La(III)-vacancy a stable electronically balanced configuration is formed. The gap of the material with the M(III) vacancy decreases from ~3.15 eV in LaSi$_3$N$_5$ to ~2.52 eV. A similar band gap is observed for both La(III) and Eu(III) vacancy and its value does not depend on the Eu(III)/La(III) substitution rate (Fig. 5, bottom lines).

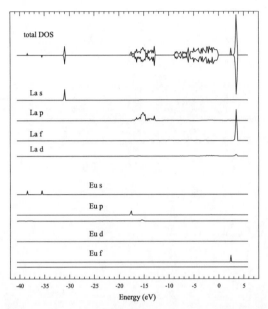

Figure 6. Total DOS of the O/N substituted and Eu(II)-doped LaSi$_3$N$_5$ and La- and Eu-projected orbital components.

Another mechanism of the compensation of the excess electron density introduced by the O-atom is the decrease of the oxidation state Eu(III)→Eu(II). The combined N/O and La(III)/Eu/(II) substitution then produces an electronically balanced configuration. The electronic structure of the O/N substituted and Eu(II)-doped LaSi$_3$N$_5$ is displayed in Figure 6. The La-projected components of the total DOS form a similar electronic structure like in the stoichiometric LaSi$_3$N$_5$ (Fig. 3). In Eu-doped compound, however, the concentration of the La d states in the bottom part of the conduction band is much lower than in LaSi$_3$N$_5$ (Fig. 6). The Eu-states exhibit a remarkable down-shift of spin-up states of both Eu s and Eu p energy levels. This asymmetry is caused by the different occupation of spin-up and spin-down states of Eu f rlrctrons. While spin-up states are completely occupied (4 f^7) spin-down states remain empty. The occupied spin-up 4f states form a band between the valence p band and the conduction band at ~2.5 eV (Fig.6). The narrowness of this band indicates a nonbonding character of the f states. The location of this band between the p band and the conduction band leads to a considerable decrease of the band gap. The DFT calculated band gap is 0.552 eV. The electronic structure indicates that upon adsorption the 4f^7 → 5d transition takes place.

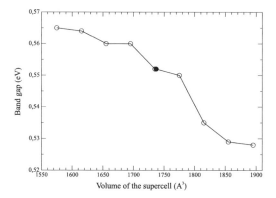

Figure 7. The band gap of Eu(II)-doped LaSi$_3$N$_5$ as a function of the cell volume (the full circle indicates the gap at the equilibrium volume).

Figure 7 shows the change of the band gap of the Eu(II)-doped LaSi$_3$N$_5$ with the variation of the cell volume. With the isomorphic change of the cell volume up to ~9% to both sides of the equilibrium volume the band gap changes only slightly between ~0.53 eV and ~0.57 eV. With the decrease of the cell volume (increase of pressure) the band gap increases, the same behavior as observed by stoichiometric LaSi$_3$N$_5$ (Fig. 4). In Eu(II)-doped material, however, the variation of the band gap with the change of the cell volume is approximately by one order smaller.

The DFT LDA or GGA approach usually fails to correctly describe strongly correlated electrons. The location of one-electron energies is such that the band gap is an order of magnitude smaller than that observed by electron spectroscopy. The origin of this failure in transition metal compounds is known to be associated with an inadequate description of the strong Coulomb repulsion between 4f electrons localized on the metal ion[16]. The method commonly utilized to estimate the band gap uses the orbital-dependent functional LDA+U. The orbital-dependent interaction is considered only for highly localized atomic-like f orbitals on the same site (Eu). The effect of the added term is to shift the localized orbitals relative to the other orbitals. The VASP code allows the application of two LDA+U approaches. The variation of the band gap of Eu(II)-doped and O/N substituted LaSi$_3$N$_5$, calculated using the simplified rotationally invariant of the LDA+U by Dudarev et al.[14], is displayed in Figure 8. The on-site electron-electron interactions are treated using U and J parameters (effective on-site Coulomb and Exchange parameters). Fig. 8 shows that using U=J=0 the band gap of the value 0.549 eV is practically identical with the standard DFT band gap. With the increase of the parameter U and J the 4f band is shifted to more negative binding energies, thus increasing the value of the band gap. At U=9 and J=0 (U=10, J=1) the 4f band is completely immersed in the p bonding band and the band gap of 3.037 eV is determined by the difference between the highest occupied one-electron energy of the p band and the lowest unoccupied energy level of the conduction band.

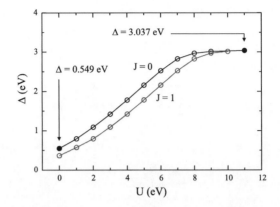

Figure 8. Variation of the band gap with parameters U and J of the LDA+U method.

The LDA+U method allows the tuning of the position of the f band relative to the position of the other bands, thus tuning the value of the band gap. Still uncorrected, however, remain unoccupied one-electron states of the conduction band. The position of these states is too low, causing a too small value of the band gap. To calculate a more precise band gap of the Eu(II)-doped and O/N substituted LaSi$_3$N$_5$ a hybrid functional is used proposed by Perdew at al.[15] in the form

$$E_{xc} = E_{xc}^{LDA} + 1/4(E_x^{HF} - E_{xc}^{GGA}).$$

The exchange-correlation energy E_{xc} is defined as a combination of the LDA exchange-correlation energy E_{xc}^{LDA}, the Hartree-Fock exchange energy E_x^{HF}, and the GGA exchange-correlation energy E_{xc}^{GGA}. Instead of proposed mixing parameter ¼ a value of 0.207 is used.

The total DOS and the La- and Eu-projected orbital components calculated using the hybrid functional are displayed in Fig. 9. The electronic structure calculated using the hybrid functional is similar to that calculated using the standard DFT (Fig. 6). For the spin-up states of the s- and the p-band of Eu a down-shift is observed, caused by the asymmetry in filling of Eu f states. While spin-up f-states are fully occupied the Eu spin-down f-states remain empty. The Eu f energies form a band of nonbonding states between the p band and the conduction band. The Eu d states are empty and located above the displayed range of energies. The hybrid-functional electronic structure show, however, several new features. The bottom of the conduction band is formed by the La 5d states, in contrast to La 4f states in the electronic structure by standard DFT (Fig. 6). Upon the excitation of the ground state electrons the Eu 4f → La 5d transition occurs. The highest occupied Eu 4f nonbonding energy levels are shifted downwards thus increasing the band gap to a value larger than 3 eV.

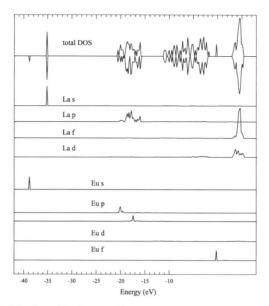

Figure 9. Total DOS of the O- and Eu(II)-doped LaSi$_3$N$_5$ and La- and Eu-projected orbital components calculated using the hybrid functional.

The band gap calculated using the hybrid functional is listed in Table 3 along with the band gap calculated using the standard DFT and the LDA+U approach. The band gap calculated using the hybrid functional of 3.1 eV reasonably compares with frequencies of the adsorption band reported for Eu-doped LaSi$_3$N$_5$ by Uheda et al.[17]. The broad adsorption band extends over frequencies 490–250 nm, corresponding to transitions 2.53 – 4.97 eV, with the adsorption maximum at ~350 nm (3.55 eV). A band gap of ~3.1 eV can be estimated in the LDA+U approach using parameters U=9, J=0. This band gap, however, corresponds to an incorrect electronic structure with the Eu 4f states shifted towards negative energies and hidden below the upper edge of the p band.

Table 3. The band gap of the Eu(II)-doped and O/N substituted LaSi$_3$N$_5$ calculated by different methods (eV)

	(p–CB)*	
Standard DFT	3.086	0.542
LDA+U	3.037	0.549–3.037
Hybrid functional	4.757	3.101

*The distance between the p band and the conduction band.

CONCLUSIONS
The electronic structure of the stoichiometric and Eu-doped LaSi$_3$N$_5$ is calculated using the periodic ab initio DFT method. A 2x1x2 supercell (144 atoms) is used to mimic the realistic concentration of Eu and O. The stoichiometric LaSi$_3$N$_5$ is an insulating material with the band gap separating fully occupied states of the p-band and unoccupied La-states composed of 4f and 5d

orbitals. The mechanism of the Eu-doping is investigated for both Eu(III) and Eu(II) cations. The Eu(III) cation, substituted for the La(III) cation, does not change the value of the band gap significantly. A decrease of the band gap occurs with the creation of the M(III) vacancy (M = La, Eu). The vacancy is compensated with O atoms substituted for framework N atoms. In the Eu(II) doped material a narrow band of nonbonding Eu 4f electrons is located in the gap between the p band and the conduction band. Although the electronic structure by standard DFT is qualitatively correct, the band gap between Eu 4f states and the CB is too small (~0.5 eV). Using the parametrized LDA+U method the position of the band of nonbonding Eu 4f states is shifted. Using LDA+U approach the DFT band gap can be continuously tuned from ~0.5 eV to ~3.1 eV. The use of the hybrid functional provides a realistic electronic structure with the band gap of ~3.1 eV and the Eu 4f → La 5d transition, in good agreement with experimental data.

REFERENCES
[1]Z. Inoue, M. Mitomo, N, Ii, A Crystallographic Study of a New Compound of Lanthanum Silicon Nitride, *J. Mater. Sci.* **15**, 2915-20 (1980).
[2]M. Woike, W. Jeitschko, Preparation and Crystal Structure of the Nitrosilicates Ln$_3$Si$_6$N$_{11}$ (Ln = La, Ce, Pr, Nd, Sm) and LnSi$_3$N$_5$ (La = Ce, Pr, Nd), *Inorg. Chem.* **34**, 5105-08 (1995).
[3]G.R. Hatfield, B. Li, W. B. Hammond, F. Reidinger, J. Yamanis, Preparation and Characterization of Lanthanum Silicon Nitride, *J. Mater. Sci.*, 25, 4032-35 (1990).
[4]R. O. Jones, O. Gunnarsson, The Density Functional Formalism, its Applications and Prospects, *Rev. Mod. Phys.* **61**, 689-46 (1989).
[5]J. P. Perdew, A. Chevary, S. H. Vosko, K. A. Jackson, M. R. Pedersen, D. J. Singh, C. Fiolhais, Atoms, Molecules, Solids, and Surfaces – Applications of the Generalized Gradient Approximation for Exchange and Correlation, *Phys. Rev. B* **46**, 6671-87 (1992).
[6]A. D. Becke, A New Mixing of Hartree-Fock and Local Density Functional Theories, *J. Chem.Phys.* **98**, 1372-77 (1993).
[7]D. Vanderbilt, Soft Self-Consistent Pseudopotentials in a Generalized Eigenvalue Formalism, *Phys.Rev. B* **41**, 7892-95 (1990).
[8]G. Kresse, J. Hafner, Ab-Initio Molecular Dynamics for Open-Shell Transition Metals, *Phys. Rev. B* **48**, 13115-18 (1993).
[9]G. Kresse, J. Hafner, Norm-Conserving and Ultrasoft Pseudopotentials for First-Row and Transition-Elements, *J. Phys. Condens. Matter.* **40**, 8245-57 (1994).
[10]G. Kresse, J. Furthmuller, Efficiency of Ab-Initio Total Energy Calculations for Metals and Semiconductors Using a Plane-Wave Basis Set, *Comput. Mater. Sci.* **6**, 15-50 (1996).
[11]G. Kresse, J. Furthmuller, Efficient Iterative Schemes for Ab-Initio Total Energy Calculations Using a Plane-Wave Basis Set, *Phys. Rev. B* **54**, 11169-86 (1996).
[12]P. E. Blochl, Projector Augmented-Wave Method, *Phys. Rev. B* **50**, 17953-79 (1994).
[13]G. Kresse, D. Joubert, From Ultrasoft Pseudopotentials to the Projector Augmented-Wave Method, *Phys. Rev. B* **59**, 1758-75 (1999).
[14]S. L. Dudarev, G. A. Boton, S. Y. Savrasov, C. J. Humphreys, A. P. Sutton, Electron-Energy Loss Spectra and the Structural Stability of Nickel Oxide: An LSDA+U Study, *Phys. Rev. B* **57**, 1505-09 (1998).
[15]J. P. Perdew, M. Ernzerhof, K. Burke, Rationale for Mixing Exact Exchange with Density Functional Approximations, *J. Chem. Phys.* **105**, 9982-85 (1996).
[16]S. Hufner, Electronic Structure of NiO and Related 3d Transition Metal Compounds, *Adv. Phys.* **43**, 183-356 (1994).
[17]K. Uheda, H. Takizawa, T. Endo, H. Yamane, M. Shimada, C. –M. Wang, M. Mitomo, Synthesis and Luminiscent Property of Eu^{3+}-Doped LaSi$_3$N$_5$ Phosphor, *J. Luminesc.*, **87-89**, 967-69 (2000).

FIRST PRINCIPLE MOLECULAR DYNAMIC SIMULATIONS OF OXYGEN PLASMA ETCHING OF ORGANOSILICATE LOW DIELECTRIC MATERIALS

Jincheng Du* and Mrunal Chaudhari
Department of Materials Science and Engineering, University of North Texas, Denton, Texas
*Corresponding author. Email: jincheng.du@unt.edu.

ABSTRACT

Plasma/surface interaction plays an important role in material processing. In the microelectronic processing, plasma etching is used to selectively remove layers of dielectric or photo resist. Atomic simulations can provide detailed mechanistic understanding of plasma/surface interactions that are usually difficult to obtain in other theoretical or experimental methods. In this work, we use Density functional Theory based first principle molecular dynamics (AIMD) simulations to investigate atmospheric etching of organosilicate low dielectric materials by using model structure for the low dielectric materials. It is found that threshold energies and reaction products strongly depend on the incident angles of (^{3}P) atomic oxygen and target atom. The lowest threshold energy for reaction for carbon removal is found to be 0.1 eV when atomic oxygen attacks silicon in a direction inclined to the Si-C bond. The reaction produces a methyl radical leaving the molecule, hence a carbon removal mechanism, and forms a Si-O bond. The simulation results agree well with experimental results and support diffusion controlled etching rate dependence and dielectric constant increases due to etching.

INTRODUCTION

Plasma has been widely used in surface modifications such as etching, erosion and sputtering. Even at low temperature, the plasma species interact strongly with the surface through chemical and physical processes that lead to changes of surface properties [1,2]. Except for material processing, understanding plasma surface interaction is also important for materials for lower earth orbitals and fusion reactor applications. The detailed atomistic mechanism of plasma surface interaction is usually very difficult to obtain by pure theory or experimental techniques. Atomistic computer simulations, especially molecular dynamics (MD) simulations, have been shown to provide detailed mechanistic understanding of surface reaction induced by plasma reaction [1,2].

Atomospheric plasma differs from conventional plasma in the sense that there are much fewer charged ionic or radical species. These charged species are usually accelerated by potential biases and collide with the surface with high kinetic energies. These high kinetic events usually dominate the surface reaction and interaction in conventional plasma. In atmospheric plasma, the high pressure prevents the high kinetic energy particles through multiple gas phase collisions.

At the same time, excited atoms or radicals usually reach the ground state due to collision. This leads to difficulties in modeling atmospheric plasma etching processes where the low kinetic energy events require accurate description of chemical bond formation and breakage that usually difficult to describe using empirical potentials. Recent development of computational power and efficient first principle simulation algorithm makes the first principle based molecular dynamics or *ab initio* molecular dynamics (AIMD) simulation possible [2-3]. In this paper, we apply AIMD to understand the ground state ^{3}P oxygen etching of organosilicate low dielectric constant materials. The simulations show that carbon abstraction happen through methyl radical removal and the creation of Si-O dangling bonds, which explains the experimentally observed decrease of carbon contents on the surface after etching and increases of dielectric constant after etching.

With ever increasing miniaturization of microelectronic devices, new materials are introduced to the traditionally silica on silicon CMOS devices [4-5]. Materials with dielectric constants lower or higher than silicon dioxide (SiO_2, with a diectric constant of 3.9) have been introduced in recent commercial devices. On the backend interconnects, materials with low dielectric constants are introduced to replace silica to decrease the cross talk noises of the metal wires. Organosilicate materials are a type of low dielectric constant materials. In organosilicate materials, methyl groups and micro or nano cavities are introduced to lower the dielectric constant. Fig. 1 shows schematic picture of the structure of low k materials. The building blocks M, D, T and Q means one, two, three or four bridging oxygen on the silicon atom with the rest of the Si-O bonds replaced by Si-CH$_3$ bonds. In pure silica, only Q groups exist. The methyl groups in the M, D, T and Q structure moieties, and the cavities contribute to the lowering of dielectric constant in the material [6].

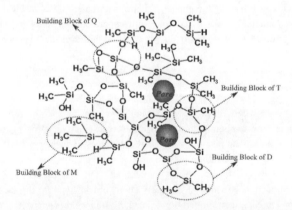

Fig. 1 Schematics of organosilicate based low dielectric materials [6].

Effective etching method is needed to selectively remove the low dielectric constant materials in microelectronic processing. Atomspheric plasma is found to be an effective method to etch organosilicate low dielectric constant materials. However, etching is also associated with increase of dielectric constant and rending the film to be more hydroscopic. Detailed understanding of oxygen plasma interaction with these materials is critical in designing etching processes and improving the dielectric film properties.

In this paper, we first introduce some details of the first principle molecular dynamics simulation method, the structure models used in the simulations, and then the observed critical collision reaction pathways are reported.

SIMULATION DETAILS

AIMD was performed using the Vienna *ab initio* Simulation Package (VASP) [7,8] on parallel implementation at Texas Advanced Computer Center (TACC) super computers. At each of the MD steps, the ground state of atomic configuration is obtained through self consistent solution of the Kohn-Sham equations, also called Born-Oppenheimer MD. In the electronic structure calculations, plane wave basis set with a kinetic energy cutoff of 400 eV was used. Projected augmented wave (PAW) pseudopotentials and generalized gradient approximation Exchange and Correlation functional with the parameter of PBE were utilized in the simulations.

The atomic oxygen was put at a distance at least 4 Angstrom away from the target atom. A velocity is then given according to the kinetic energy provided. The time step for MD simulations was varied, in the range of 0.1 to 1 femtosecond (*fm*), depending on the kinetic energies on the atomic oxygen to ensure reasonable energy conservation. Canonical ensemble (constant energy, number of atoms, and volume NVE), ie no thermostat or velocity scaling applied, was used in all the MD simulations. The initial tetramethylsilane and tetramethyloxysiloxane were fully optimized within DFT with the same basis set and energy cutoffs. The simulations were stopped when the product states are stabilized and usually last 500 to 1000 MD steps (some of the MD movies are provided as auxiliary data).

Special considerations of simulation of collision reaction atomic oxygen are also worth mentioning. The ground state of atomic oxygen is a triplet state with two unpaired electrons (^{3}P) while the first exited state is singlet state without unpaired electrons (^{1}D). In atmospheric oxygen plasma, due to large collision rates among plasma species, ground state oxygen (^{3}P) should be the dominant oxygen species. It is also important to ensure the initial state of oxygen is in the triplet state in the simulation. Due to triplet state of oxygen and the fact that we are studying bond breaking and formation, spin polarized (or open shell) *ab initio* calculations are required in the simulations although these calculations are computationally more expensive than spin non-polarized (or closed shell simulations).

RESULTS AND DISCUSSION

1. Model structures for low dielectric constant materials

As we can see from Fig. 1 that organosilicate materials are amorphous in nature. In order to apply first principle molecular dynamics simulations to study atomic oxygen collision reaction with these materials, we have to use model structures. Two model structures were chosen to represent the organosilicate materials. The first one is tetramethylsilane (TMS) which is shown in Fig. 1 (a). TMS has four Si-C bonds and four methyl groups so is a good representative structure of the organic part of organosilicate materials. The second one is the trimethylcyclotrisiloxane (TMCTS) which is shown in Fig. 2 (b). In the TMCTS macromolecule structure, there exist Si-O-Si bonds, Si-C bonds, C-H bonds and Si-H bonds. So the TMCTS structure contains all the essential bonds in an organosilicate low dielectric constant materials and thus is a very good reprentative structure to investigate plasma / organosilicate interactions. It is also worth mentioning that similar molecules of TMCTS have been used as source materials to synthesize organosilicate low dielectric constant materials using plasma synthesis methods.

The representative structures were first fully relaxed using DFT with GGA PBE functional. For the TMS structure, two types of conformation of the methyl groups were studied. The staggered structure, ie the methyl groups are 60 degree from the Si-C bonds as shown in Fig. 2 (a), was found to have a lower energy (by 0.12 eV) than the eclipsed structure and hence was used in all the simulations.

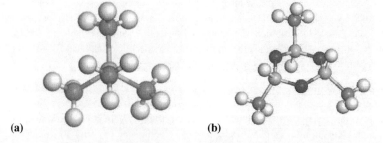

(a)   (b)

Fig. 2 Tetramethyl silane (a) and trimethyl cyclictrisilane (b) are model structures.

2. Atomic Oxygen Reaction pathways for tetramethylsilane (TMS)

Collision reaction along the Si-C bond for TMS was found to have very high kinetic energy barrier (around 3 eV), which is defined to be the lowest energy required for a collision reaction to happen. The reaction produces a formaldehyde molecule leaving the TMS and a resulting Si-H bond.

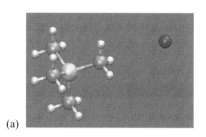
(a)

(b)

Fig. 4 Initial (a) and product (b) states of collision reaction of atomic oxygen along the Si-C bond of tetramethylsilane.

The reaction energy barrier was found to be much lower for the collision reaction with an incident angle perpendicular to the Si-C bond (around 1 eV). The reaction products is a methyl radical leaving the molecule and the formation of a Si-O bond. The reaction happens by first pushing the methyl group with the targeting C aside and then the oxygen atom forms a bond with silicon. The formation of the Si-O bond leads the scission of another Si-C bond (not the one that was targeted by atomic oxygen) the formation of a methyl radical. This reaction leads to us to think the low kinetic energy collision reactions could be associated with Si-O bond formation. This turns out to be critical in finding low energy reaction pathways for atomospheric oxygen plasma reaction since the room temperature plasma has a very low kinetic energy.

(a)   (b)

Fig. 5 Initial (a) and product (b) states of atomic oxygen collision reaction along the direction perpendicular to the Si-C bond.

3. Atomic Oxygen Reaction pathways for trimethylcyclictrisiloxane (TMCTS)

The reactions pathways for atomic oxygen collision reaction with TMCTS have been studied and the threshold energies were found for each of these reactions. The reaction direction, threshold kinetic energy, and reaction product states are summarized in Table 1 and the corresponding atom number is shown in Fig. 6 [2]. The reaction mechanisms shown the energy barrier can vary considerably, e.g. the threshold energy for collinear collision (mechanism 2) along Si-C bond is 3 eV while several others such as inclined attach of Si to the Si-C bond direction has a much lower threshold energy of 0.1 eV (mechanism 1). Also, the product states vary considerably from carbon abstraction through methyl formation, hydrogen abstraction through OH or H radical formation or through water molecule formation. Most importantly, the reaction threshold energy and product states depend on the target atom and the collision direction. This is clearly shown in the listed reaction mechanisms in Table 1.

Table 1 Summary of AIMD simulations for $O(^3P)$ interaction with TMCTS. Atom numbers are shown in Fig. 6.

Reaction Mechanism number	Targeting Atom	Incident Direction	Threshold Kinetic Energies (eV)	Product states
1	Si2	Inclined to Si2-C2	0.1	CH_3+Si-O
2	C1	Collinear Si1-C1	3	H+Si-COH$_2$
3	C1	Side, perpendicular to Si1-C1	1	CH_3+Si-OH
4	C1	Side, perpendicular to Si1-C1	0.25	OH (radical)
5	Si1	Direct on Si1	0.1	H$_2$O
6	H (_Si)	Along Si-H bond	0.1	OH (radical)
7	H(_C)	Along C-H bond	0.1	OH (radical)

We are mostly interested to find the reaction pathway that can lead to carbon abstraction and happens at low kinetic energies. The reason is that in room temperature atmospheric plasma, the available kinetic energy on ground state (^3P) atomic oxygen is very low and it has been found experimentally that oxygen plasma etching leads to carbon loss of the dielectric films. The lowest reaction pathway for carbon abstraction was found to happen in the inclined attach on the silicon atom which results in methyl radical formation and Si-O dangling bonds (mechanism 1). The collision reaction initial and product states are shown in Fig. 6.

The lowest threshold energy reaction mechanism for methyl abstraction only happens when the incident oxygen attack silicon through the inclined direction relative to the Si-C bonds. This high selectivity of reaction direction suggests that atomic oxygen can withstand many collisions prior to the specific reaction before it finds the right attacking angle. The pathways demonstrate that thermal O (^3P) abstraction of CH₃ groups from an organosilicate matrix can happen at low temperature. The sensitivity of the energy barrier height to the O (^3P) trajectory and the specific target atom provide a rationale for the reported diffusion-dominated kinetics of CH₃ abstraction [5]. Further, the general agreement of the AIMD results with experimental data demonstrates the potential of first principle based molecular dynamics simulations, when combined with carefully chosen model systems, for low energy radical-induced reactions, allowing future predictive modeling of such processes.

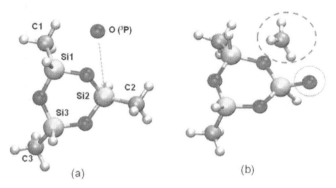

Fig. 6 [3]P atomic oxygen collision reaction with TMCTS. The inclined attach on silicon has the lowest reaction energy barrier (0.1 eV). The reaction product state is the formation of a Si-O bond and a methyl radical leaving the TMCTS molecule [2].

CONCLUSIONS

By using first principal molecular dynamics simulations, we have investigated ground state (^{3}P) atomic oxygen collision reaction with structure models for low dialectic materials. It is found that the reaction threshold energy and product states highly depend on the target atom and collision direction. Low threshold energy pathways for carbon abstraction have been found that help to explain experimental results to rationalize a diffusion-dominated etching behavior. The results show the great potential of first principle molecular dynamics simulations in the study of plasma surface interactions, which can eventually lead to predictive modeling of the complex plasma processes.

REFERENCES

1. D. B. Graves and P. Brault, J. Phys. D. Appl. Phys. 42 194011 (2009).
2. M. Chaudhari, J. Du, S. Behera, S. Manandhar, S. Gaddam, J. Kelber, Appl. Phys. Lett, 94, 204102 (2009).
3. D. Troya and G. Shartz, J. Chem. Phys. 120, 7696 (2004).
4. A. Grill and D. A. Neumayer, J. Appl. Phys. 94, 6697 (2003).
5. J. Bao, L. Shi, J. Liu, H. Huang et al, J. Vac. Sci. Technol. 26, 219 (2008).
6. C. A. Yuan, O. van der Sluis, W. D. van Driel, G. Q. Zhang, Microelect. Reliab. 48 833 (2008).
7. G. Kresse and J. Furthmuller, Compt. Mater. Sci. 6 15 (1996).
8. G. Kresse and J. Furthmuller, Phys. Rev. B 54 11169 (1996).

KINETIC MONTE CARLO SIMULATION OF CATION DIFFUSION IN YTTRIA-STABILIZED
ZIRCONIA

Brian Good
Materials and Structures Division
NASA GRC
Cleveland, Ohio

ABSTRACT
Yttria-stabilized zirconia (YSZ) is of interest to the aerospace community, notably for its application as
a thermal barrier coating. In such an application, the inhibition of oxygen diffusion is a major concern,
and this issue has been addressed via molecular dynamics and kinetic Monte Carlo simulation.
However, the mechanical integrity of such coatings can be affected by processes involving diffusive
motion on the cation lattice.

In this work, we perform kinetic Monte Carlo simulations to investigate cation diffusion in YSZ. Using
diffusive migration barrier energies from the literature, we obtain diffusivities that are several orders of
magnitude smaller than oxygen diffusivities from experiment, or from earlier simulations. We report
on the effects on cation diffusivity of cation sublattice vacancy concentration, Y concentration, and
temperature, and relate these results to oxygen diffusion in the same materials.

INTRODUCTION
Zirconia-based materials are of interest for a variety of technological applications, ranging from fuel
cell electrolytes to thermal barrier coatings for turbine engine components. In some applications, the
high diffusive ionic oxygen conductivity is a desirable feature, while for others a large oxygen
diffusivity is problematic. In addition, the high-temperature phases of zirconia are not stable at room
temperature, limiting its usefulness in applications involving significant thermal cycling.

Pure zirconia exists in a monoclinic structure below about 1100C, a tetragonal structure between
1100C and 2300C, and a cubic structure between 2300C and the melting point [2]. The existence of
such phase transitions can limit the utility of zirconia for high-temperature applications.

However, substitutional cation doping with aliovalent ions such as Ca^{2+} or Y^{3+} can stabilize the
tetragonal or cubic phases. In such cases, additional oxygen vacancies are formed so as to maintain
electrical neutrality, which tends to enhance the oxygen ionic conductivity. This makes such materials
of interest for use as oxygen sensors, or as solid electrolytes for fuel cells.

In particular, the presence of yttrium on the fcc cation sublattice can affect the oxygen diffusivity; at
low Y concentrations the diffusivity increases with Y concentration, reaches a maximum between 8
and 15 mol%, and decreases at higher Y concentrations [1].

On the other hand, when stabilized through the addition of Y_2O_3, the cubic form remains stable over a
very wide range of temperatures, from room temperature to the melting point. The low thermal
conductivity and high thermal stability of such materials make them potentially useful in high-
temperature applications, e.g. thermal barrier coatings for jet turbine blades. In such applications, high
oxygen ionic conductivity is not desirable.

127

The material of interest here, yttria-stabilized zirconia (YSZ), exists in a cubic flourite structure in which the Zr and Y cations are located at sites on a face-centered cubic sublattice, while the oxygen ions are located in a simple cubic sublattice whose lattice constant is one-half that of the cation sublattice. Oxygen diffusion takes place primarily via the hopping of oxygen ions to nearest-neighbor oxygen vacancy sites on the oxygen sublattice.

In this work we investigate cation diffusion in yttria-stabilized zirconia (YSZ) using a kinetic Monte Carlo (kMC) computer simulation procedure. We discuss the dependence of cation diffusivity on cation concentration, cation and oxygen vacancy concentration, and temperature, and relate these results to oxygen diffusion in the same materials.

Because of the range of potential applications, considerable experimental and theoretical effort has gone into understanding the behavior of YSZ, and, in particular, oxygen diffusion in YSZ and related materials. Of interest here, computer simulations using a variety of techniques have been performed.

Schelling et al. [3] investigate the cubic-to-tetragonal phase transition via molecular dynamics simulation, and correctly predict the experimentally observed stabilization due to yttrium doping of ZrO_2. Similar work has been carried out by Fabris et al. [4]. Fevre et al. investigate the thermal conductivity of YSZ using Monte Carlo [5] and molecular dynamics [6] techniques.

Krishnamurthy et al. [7,8] have performed kinetic Monte Carlo simulations of oxygen diffusion in YSZ. They describe the Y-dependence of the diffusivity as being due to a competition between two mechanisms.

First, increasing the Y concentration leads to an increase in vacancies on the oxygen sublattice; each pair of substitutional Y cations requires the formation of one oxygen vacancy in order to maintain electrical neutrality, thus increasing the number of sites available for oxygen ion hopping conduction.

Second, an oxygen ion hopping to an adjacent oxygen vacancy site must pass between two cations, as shown in Figure 1. The energy barrier that governs such hopping is strongly dependent on the chemical species of the barrier cation. In particular, the barrier energy for a hop between two Zr cations (typically about 0.5eV as computed using density functional theory), is substantially smaller than the energy for a hop between a Zr and a Y ion (typically 1.3eV), which in turn is substantially smaller than the barrier for a hop between two Y ions, typically 1.8-2.0eV. Therefore, increasing the Y ion concentration leads to an increase in the numbers of Zr-Y and Y-Y barrier pairs, which tends to inhibit oxygen ionic conductivity. The competition between these two mechanisms gives the observed cation concentration dependence of the oxygen conductivity in YSZ.

A number of other computational studies of oxygen diffusion in YSZ have been performed, including molecular dynamics studies by Kahn et al. [9], Okazaki et al. [10], Perumal et al. [11] and Shimojo et al. [12].

While oxygen diffusion in YSZ is quite rapid, there is also a much slower cation diffusion that can lead to adverse effects on material structure via creep.

Kilo et al. [13] analyze Zr diffusion data from creep data, dislocation loop shrinkage data, and Zr tracer diffusion data to identify the defect(s) responsible for cation diffusion. They identify diffusion involving the single cation vacancy as the most likely mechanism. However, measurements of activation enthalpies remain problematic; the relative ordering of the enthalpies for Zr and Y diffusion are not consistent among various studies.

Kilo et al. [14] perform simulations of cation diffusion in YSZ (as well as doped lanthanum gallates) via molecular dynamics, using NPT molecular dynamics and a Buckingham + Coulomb potential. They consider the hopping of cations via vacancy sites, introduced in the form of Schottky defects, at a concentration of 0.004, for yttria concentrations of 11, 19 and 31 mol %.

They find that the diffusion coefficients for Y and Zr are significantly different, with Y diffusion 3-5 times faster than Zr diffusion. They report calculated enthalpies, for 11 mol % Y_2O_3 YSZ, of 4.8eV (Y) and 4.7eV (Zr). These differ in both magnitude and ordering from experimental results from Kilo et al. [13], who report 4.6eV (Y) and 4.2eV (Zr). The molecular dynamics results also show that cation diffusivities are independent of Y_2O_3 concentration, or slightly increasing with increasing Y_2O_3 concentration, in contrast with experiment, which shows cation diffusivity to decrease with increasing Y_2O_3 concentration.

Figure 1. Oxygen (left) and cation (right) diffusion paths in YSZ. Cation-grey; Oxygen-red; Oxygen or cation target vacancy-green.

KINETIC MONTE CARLO METHOD
The kinetic Monte Carlo method is a simulation procedure that mimics the dynamical evolution of a system using a stochastically generated sequence of state-to-state transitions. While the method does not produce the detailed atomistic trajectories available from Molecular Dynamics (MD) simulation, it can produce state-to-state trajectories that are, in principle, exact. It differs from the more widely used Metropolis Monte Carlo procedure, which is commonly used to study equilibrium properties rather than dynamics. Metropolis Monte Carlo produces ensemble averages of equilibrium physical

properties of a system from a sequence of random sample configurations of the system, chosen so as to be consistent with, e.g., the Boltzmann distribution.

In particular, kMC is useful in the study of processes involving so-called infrequent events. Simulating such processes via MD can be problematic, especially if the temperature is low and the events of interest are thermally activated. An MD simulation of such a system reproduces the detailed dynamics of the events, separated by long periods during which no such events are observed. Consequently, MD simulation of such processes is often inefficient and computationally expensive.

KMC, on the other hand, incorporates a detailed treatment of the events of interest, but a less detailed (and computationally less expensive) treatment of the intervals between events. As such, the method provides access to dynamical processes that take place over longer time scales than are accessible via MD. It should be noted, however, that if different processes in a kMC simulation have event probabilities that differ by many orders of magnitude, such simulations may become quite inefficient.

Each step of a kMC simulation requires an enumeration of all possible events, each defined in terms of an initial and final state of the system, along with the transitional state that determines the transition rate. For each event, a transition probability is calculated, and a catalog of all events and the corresponding transition probabilities is constructed. One of the events is chosen from the catalog stochastically, and the system is adjusted to reflect the occurrence of the event. After each event, the catalog is updated to reflect the new state of the system, and the simulation clock is advanced by a stochastically chosen time interval.

Transition probabilities for such simulations are often derived from harmonic transition state theory. The result of interest here is that, if the energy barrier E associated with an even is known, the transition rate at temperature T is proportional to $\exp(-E/k_BT)$ times a prefactor that depends on the normal mode frequencies at the energy minimum and at the saddle point associated with the transition.

In the case of vacancy-modulated diffusive motion, the state of the system is characterized by the distribution of chemical species and vacancies within the computational cell under study. In such a system, a state-to-state transition involves a thermally activated hop of an atom to a vacancy site. These transitions are infrequent; an atom in the vicinity of a vacancy undergoes thermal oscillation around the local potential minimum, but only rarely undergoes a hop to the neighboring vacancy site. The probability per unit time that an atom experiences such a hop depends primarily on the height of the energy barrier, which in turn depends on the configuration of the atoms in the vicinity of the hopping atom at its pre-hop position, and at the saddle point during the hop.

We model YSZ using a cubic computational cell consisting of 6144 atoms, incorporating periodic boundary conditions. In the absence of cation vacancies, the number of oxygen vacancies is obtained from the requirement that the computational cell remain electrically neutral. When cation vacancies are included, we modify the number of oxygen vacancies so that the cell remains neutral; however, the cation vacancy concentration is typically small, and we restrict our attention primarily to cation vacancy concentrations less or equal to than one percent. It may be the case that oxygen vacancies may preferentially occupy sites near one of the cation species, but we neglect this effect, and distribute the oxygen vacancies randomly within the computational cell. Cation vacancies are also initially randomly distributed.

Based on preliminary unrelaxed ab initio calculations, cation hops in the [110] directions to nearest-neighbor cation vacancy sites are energetically unfavorable; barrier energies in the [100] directions are lower by about a factor of two. Given the size of the cations and the relatively tight packing of the crystal structure, we ignore hopping via interstitial sites and consider only [100] hops.

When the migration barrier energies are known, the cation hopping rates may be computed:

$$v_{AB} = v^0 \exp(-E_{AB}/k_B T)$$

where v_{AB} and E_{AB} are the hopping rate and barrier energy for a hop between cation sites A and B respectively, and v_0 is the frequency factor. v_0 is typically assigned a value between 10^{12} and 10^{13} for these materials; given that the measured cation diffusivities from different experimenters can differ by as much as an order of magnitude, we assume a value of 10^{13} with the understanding that the values of the diffusivities presented here involve considerable uncertainty. Given the similarity of the atomic masses, it is not expected that the frequency factors for Zr and Y will differ appreciably.

In these simulations, each cation vacancy gives rise to a number of possible hops from its [100] neighbors. For each possible hop, the hopping rate and corresponding hopping probability are computed, where the hopping probability is given by

$$P_{AB} = \frac{v_{AB}}{\nu}$$

where ν is the sum of hopping rates for all possible hops in the computational cell. A catalog of all possible hops, and the corresponding hopping rates, is created.

During the kMC process, one of the possible hops is chosen probabilistically from the catalog, and the hopping cation is transferred to the target vacancy site. Hopping rates for all possible hops involving the new vacancy location are computed and added to the catalog, while rates involving the vacancy's previous location are deleted, and the sum of the hopping probabilities is updated. Finally, the simulation clock is advanced by a stochastically chosen timestep:

$$\Delta t = -\frac{\ln(R)}{\nu}$$

where R is a random number greater than zero and less than or equal to unity.

When the simulation has run long enough to accumulate statistically useful information, the mean square displacement, averaged over all vacancies, is computed. The vacancy diffusivity D is obtained from the Einstein relation

$$\langle R^2 \rangle = 6D t$$

and the cation diffusivity D_c is obtained by balancing the number of vacancy and cation hops:

$$D_c = \frac{C_{CV}}{1 - C_{CV}} D$$

where C_{CV} is the concentration of cation vacancies.

RESULTS AND DISCUSSION

In all simulations performed to date, only two distinct types of hops are considered: Y and Zr hops along a [100] direction to a vacancy one lattice constant away. This restriction involves three assumptions. First, it neglects the fact that there are two sets of oxygen cations that function as barriers for such hops, as shown in Figure 1. These sets consist of four oxygen ions arranged in a square, located 1/4 and 3/4 of the way long the hopping path. We ignore the possibility that a cation may hop to an interstitial site halfway along the path, and at a later time continue on to the target vacancy site, or return to its initial location. The interstitial site is less favorable energetically than are the on-lattice cation sites, and we therefore consider the hop to be a single event, regardless of whether it is in fact a concerted event.

Second, we neglect the fact that the four oxygen barrier sites need not be fully populated. Experimental results do not provide this level of detail, but preliminary unrelaxed ab initio calculations suggest that the energy barrier for a hop across a barrier consisting of three oxygen ions and an oxygen vacancy is about ten percent lower than that of a hop across a barrier fully populated with oxygen ions. In view of the small size of this difference, we neglect the possibility and assume only fully populated barriers.

Third, we neglect barrier energy differences due to variation in the species of the cation nearest neighbors of the hopping cation and target vacancy, and assume a single hopping energy barrier for each cation species. More detailed ab initio energy barrier calculations that resolve these differences are currently under way.

KMC simulations have been performed for a range of compositions, with Y_2O_3 mole fractions ranging from 0.01 to 0.40. A range of cation vacancies is considered, and the concentration of oxygen vacancies is adjusted in each case to guarantee cell neutrality.

Cation diffusivity is very sensitive to the choice of diffusive barrier energy. However, the values of those quantities have not been definitively established. Solmon et al [15] report hopping enthalpies of 4.8-4.95eV in the range of 1300-1700C. Gomez-Garcia et al [16] report enthalpies of 5.5-6.0eV above 1500C. Chien and Heuer [17] report a value of 5.3eV at 1100-1300C, and Dimos and Kohlstedt [18] find a value of 5.85eV at 1400-1600C. Kilo et al [13] find values of 4.4-4.8eV at 1125-1460C. Mackrodt et al [19] have calculated Y and Zr migration energies of 2-7eV for YSZ.

Given the wide range of energies, there is some latitude in choosing the values for our simulations; we have chosen enthalpies of 3.7 and 3.62 for Y and Zr migration, respectively. These values have been chosen to produce diffusivities within the range of experimental values at 2000K, though the diffusivities are toward the low end of reported experimental values. The Y and Zr values are different in the same proportion as values reported by Kilo et al. [14].

We are currently calculating migration energies using ab initio density functional theory. The unrelaxed values of these energy barriers are too large, and produce diffusivities that are substantially too small, but fully relaxed energies are not yet available.

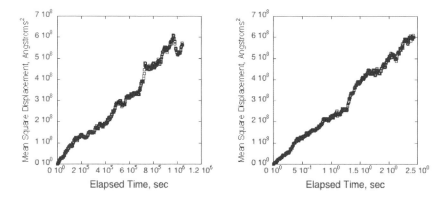

Figure 2. Mean square diffusive distance as a function of simulation time. 1500K (left), 2750K (right).

The kMC process produces vacancy trajectories that are random walks, although step probabilities depend on the species of the hopping cation. The mean square displacement of such walkers is expected to be linear in time, and that behavior is qualitatively displayed in Figure 2.

As previously mentioned, earlier studies of oxygen diffusion in YSZ, using similar methodology [7,8] found that the behavior of the diffusivity as a function of Y concentration was the result of a competition between the increased oxygen vacancy concentration that corresponds to an increase in Y concentration, and the increase in higher-energy hopping energy barriers due to the increased presence of Zr-Y and Y-Y barrier pairs.

The situation for cation diffusivity is somewhat different. The oxygen vacancy concentration is again tied to the concentrations of Y ions as well as cation vacancies; in both cases, substituting Y^{3+} cations or cation vacancies for Zr^{4+} vacancies requires additional oxygen vacancies to be created.

As the cation vacancy concentration increases, one expects the cation diffusivity to increase as a consequence, as there are more available vacancy sites to function as targets for hopping cations. In addition, the corresponding increase in oxygen vacancies makes it increasingly likely that one or more of the barrier oxygen atom sites may contain a vacancy. Initial ab initio DFT calculations suggest that replacing one of the four barrier oxygen sites lowers the energy barrier on the order of ten percent, which tends to increase the diffusivity. Because this difference is relatively small, we expect the dependence of the diffusivity on cation vacancy concentration to be dominated by the increase in available hopping sites, with the result that the diffusivity should increase strongly with cation vacancy concentration. This behavior is exhibited by our kMC simulations, as shown in Figure 3, where the cation diffusivity is seen to increase linearly with cation vacancy concentration.

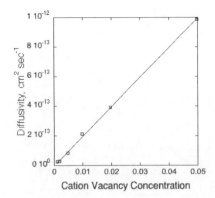

Figure 3. Cation diffusivity versus cation vacancy concentration.

The dependence of cation diffusivity on Y^{3+} concentration is more complicated, and results of other researchers are ambiguous. As the Y concentration increases, there are two effects on cation diffusivity. Increasing the Y cation concentration also increases the number of oxygen vacancies, resulting in a greater number of barrier oxygen complexes that contain one or more vacancies, which tends to weakly enhance the diffusivity. Increasing the Y concentration also increases the ratio of Y to

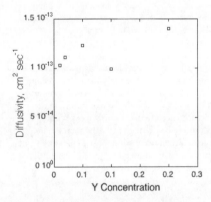

Figure 4. Dependence of cation diffusivity on Y cation concentration.

Zr cation hops, and the effect of this on the diffusivity is not clear due to the uncertainty in the barrier energies, and their ordering. Regardless, it is likely that the difference in energy barriers for Zr and Y hops is relatively small (unrelaxed ab inito calculations suggest differences on the order of ten percent), so that the effect on diffusivity is probably small. Depending on the relative sizes of the barriers, increasing the Y concentration may have either the same, or the opposite, effect as increasing the vacancy concentration, so that either a weak increase or a weak decrease in diffusivity with increasing Y concentration is plausible. KMC results are shown in Figure 4; there is considerable. scatter in the data, but it appears that there is a weak increase with increasing Y concentration. This trend is consistent with that reported by Kilo et al. [14] from molecular dynamics simulations, but disagrees with experiment. Given that experimental and theoretical results from other workers show inconsistent dependence on Y concentration, we consider this an open question.

The temperature dependence of cation diffusivity is shown in Figure 5, along with experimental results. Because the diffusive energy barriers were chosen to duplicate one experimental result at 2000C, the agreement between this work and experiment is not unexpected. The slope of the line from our kMC calculations is less than that of the experimental results shown, although it differs from the results of Chien et al. by less than ten percent. Other kMC simulation results (not shown) using larger

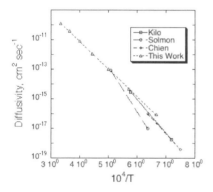

Figure 5. Temperature dependence of cation diffusivity.

barrier energies show a slope more consistent with other experimental results, although, as described above, the absolute values of the diffusivity are smaller. In any event, the reasonable agreement between simulation and experiment of the slopes of the $\ln(D)$ versus $1/T$ suggests that the kMC simulations capture the fundamentals of the diffusion process.

CONCLUSIONS AND OUTLOOK
We have performed kinetic Monte Carlo computer simulations of cation diffusion in yttria stabilized zirconia. Cation diffusivities computed here are several orders of magnitude lower than oxygen diffusivity in the same materials, qualitatively consistent with experimental observation, and with molecular dynamics results. In order to produce diffusivities consistent with experiment, we have used

hopping barrier energies that are within the range of experimental values, but smaller than most experimental numbers. The temperature dependence of the kMC results is in reasonable agreement with experiment, as is the qualitative dependence on cation vacancy concentration. The diffusivity increases weakly with Y concentration, a result that qualitatively agrees with results from molecular dynamic simulation but disagrees with experimental results.

The kMC model is currently undergoing revision in an attempt to improve its fidelity. New barrier energies are being computed, using ab initio density functional theory, for a variety of cation and oxygen barrier configurations. These results will allow us to more realistically take into account the random distribution of oxygen and cation vacancies, and cation dopants. In addition, we are developing a lattice Monte Carlo code to determine whether, at the level of fidelity of our current work, oxygen vacancies exhibit a preference for Zr or Y neighbors.

REFERENCES
[1] R. E. W. Casselton, Phys. Status Solidi A, 2, 571-585 (1970).

[2] P. Aldebert and J. P. Traverse, J. Am Ceram. Soc. 68 [1], 34-40 (1985).

[3] P. K. Schelling, S. R. Phillpot and D Wolf, J. Am Ceram. Soc. 84 [7], 1609-1619 (2001).

[4] S. Fabris, A. T. Paxton and M. W. Finnis, Phys. Rev. B 63, 094101 (2001).

[5] M Fevre, A. Finel and R. Caudron, Phys. Rev. B 72, 104117 (2005).

[6] M Fevre, A. Finel, R. Caudron and R. Mevrel, Phys. Rev. B 72, 104118 (2005).

[7] R. Krishnamurthy, Y.-G. Yoon, D. J. Srolovitz and R. Car, J. Am. Ceram. Soc. 87 [10],1821-1830 (2004).

[8] R. Krishnamurthy, D. J. Srolovitz, K. N. Kudin and R. Car, J. Am. Ceram. Soc. 88 [8],2143-2151 (2005).

[9] M. S. Kahn, M. S. Islam and D. R. Bates, J. Mater. Chem. 8 [10], 2299-2307 1998.

[10] H. Okazaki, H. Suzuki and K. Ihata, Phys. Let. A 188, 291-295 (1994).

[11] T. P. Perumal, V. Sridhar, K. P. N. Murthy, K. S. Easwarakumar and S. Ramasamy, Comp. Mat. Sci. 38, 865-872 (2007).

[12] F. Shimojo, T. Okabe, F. Tachibana, M. Kobayashi and H. Okazaki, J. Phys. Soc. Jpn 61, 2848-2857 (1992), and F. Shimojo and H. Okazaki, J. Phys. Soc. Jpn 61, 4106-4118 (1992).

[13] M. Kilo, G. Borchardt, C. Lesage, O. Kaitsov, S. Weber and S. Scherer, J. Eur. Ceram. Soc., Faraday Trans. 5, 2069 (2000).

[14] M. Kilo, M. A. Taylor, C. Argirusis, G. Borchardt, R. A. Jackson, O Schulz, M. Martin and M. Weller, Solid State Ionics 175, 823-827 (2004).

[15] H. Solmon, C. Monty, M. Filial, G. Petot-Ervas and C. Petot, Solid State Phenom. 41, 103 (1995).

[16] D. Gomez-Garcia, J. Martinex-Fernandez, A. Dominguez-Rodriguez and J. Castaing, J. Am. Ceram. Soc. 80, 1668-1672 (1997).

[17] F. R. Chien and A. H. Heuer, Philos. Mag. A 73, 681-697 (1996).

[18] D. Dimos and D. L. Kohlstedt, J. Am. Ceram. Soc. 70, 277 (1987).

[19] W. C. Mackrodt and P. M. Woodrow, J. Am. Ceram. Soc. 68, 277 (1986).

DYNAMIC NEUTRON DIFFRACTION STUDY OF THERMAL STABILITY AND SELF-RECOVERY IN ALUMINIUM TITANATE

I.M. Low and Z. Oo
Department of Imaging & Applied Physics, Curtin University of Technology, GPO Box U1987, Perth, WA 6845, Australia

ABSTRACT

Aluminium titanate (Al_2TiO_5) is an excellent refractory and thermal shock resistant material due to its relatively low thermal expansion coefficient and high melting point. However, Al_2TiO_5 unstable and undergoes a eutectoid-like decomposition to α-Al_2O_3 and TiO_2 (rutile) at the temperature range of 900-1280°C. In this paper, we describe the use of high-temperature neutron diffraction to study (a) the phenomenon of self-recovery in decomposed Al_2TiO_5, and (b) the role of grain size on the rate of isothermal decomposition at 1100°C. It is shown that the process of decomposition in Al_2TiO_5 is reversible whereby self-recovery occurs readily when decomposed Al_2TiO_5 is re-heated above 1300°C, and the rate of phase decomposition increases as the grain size decreases.

INTRODUCTION

Aluminium titanate (Al_2TiO_5) is an excellent refractory and thermal shock resistant material due to its relatively low thermal expansion coefficient ($\sim 1 \times 10^{-6}$ °C^{-1}) and high melting point (1860°C). It is one of several materials which is isomorphous with the mineral pseudobrookite (Fe_2TiO_5).[1,2] In this structure, each Al^{3+} or Ti^{4+} cation is surrounded by six oxygen ions forming distorted oxygen octahedra. These AlO_6 or TiO_6 octahedra form (001) oriented double chains weakly bonded by shared edges. This structural feature is responsible for the strong thermal expansion anisotropy which generates localised internal stresses to cause severe microcracking. Although this microcracking weakens the material, it imparts a desirable low thermal expansion coefficient and an excellent thermal shock resistance.

At elevated temperature, Al_2TiO_5 is only thermodynamically stable above 1280°C and undergoes a eutectoid-like decomposition to α-Al_2O_3 and TiO_2 (rutile) within the temperature range 900-1280°C.[3-11] This undesirable decomposition has limited its wider application. Hitherto, the role of grain size on the rate of phase decomposition is poorly understood but experimental evidences suggest a nucleation and growth controlled process. It is generally agreed that the decomposition rate peaks at 1100°C and that residual alumina particles might act as preferred nucleation sites for the decomposition.[3]

In recent studies by Low and co-workers,[12-15] microstructure and furnace atmosphere have been observed to have a profound influence on the thermal stability of Al_2TiO_5. For instance, the decomposition rate of Al_2TiO_5 at 1100°C is significantly enhanced in vacuum (10^{-4} torr) or argon where >90% of Al_2TiO_5 decomposed after only 4 h soaking when compared to less than 10% in atmospheric air.[12-14] This suggests that the process of decomposition of Al_2TiO_5 is susceptible to environmental attack or sensitive to the variations in the oxygen partial pressure during ageing. The stark contrast in the mechanism of phase decomposition is believed to arise from the vast differences in the oxygen partial pressure that exists between air and vacuum.

A similar phenomenon, although less profound, has been observed for Al_2TiO_5 with a distinct difference in grain size. However, it is unclear whether there is a critical grain size associated with this phenomenon. The reason for this grain-size effect is unclear at this stage although it may be closely related to its greater tendency for microcracking as the grain size increases. The microcracking phenomenon is closely related to the material microstructure and thermal expansion anisotropy.[16-18] Below a critical grain size, the elastic energy of the system is insufficient to nucleate microcracks during cooling and thus causing no degradation to the mechanical strength. The density of microcracks increases drastically with grain size once the critical value is exceeded.

In this paper, we present results on the role of grain size on the isothermal stability of Al_2TiO_5 at 1100°C as well as its capability to self-recover when it is reheated following decomposition. The temperature-dependent thermal stability and isothermal decomposition of Al_2TiO_5 have been dynamically monitored and characterized using neutron diffraction to study the structural changes occurring during phase decomposition in real time.

EXPERIMENTAL METHODS
Sample Preparation
The starting powders used for the synthesis of Al_2TiO_5 (AT) consisted of high purity commercial alumina (99.9% Al_2O_3) and rutile (99.5% TiO_2). One mole of alumina powder and one mole of rutile powder were initially mixed using a mortar and pestle. The powder mixture was then wet mixed in ethanol using a Turbula mixer for 2.0 h. The slurry was dried in a ventilated oven at 100°C for 24 h. The dried powder was uniaxially-pressed in a steel die at 150 MPa to form cylindrical bars of length 20 mm and diameter 15 mm, followed by sintering in a air-ventilated furnace at (a) 1400°C in air for 1 h to achieve a fine-grained (FG) microstructure (~1-3µm); (b) 1500°C in air for 2 h to achieve a medium-grained (MG) microstructure (~5-10µm), and (c) 1600°C in air for 4 h to achieve coarse-grained (CG) (~30-50µm) Al_2TiO_5.

Neutron Diffraction (ND)
A medium resolution powder diffractometer (MRPD) located at the Australian Nuclear Science and Technology Organization (ANSTO) in Lucas Heights, NSW was used for neutron diffraction study of the thermal stability of Al_2TiO_5. The effect of grain size on the isothermal stability of Al_2TiO_5 was dynamically monitored at 1100°C in air atmosphere for up to 10 hours. A decomposed medium-grained Al_2TiO_5 sample was used for the study of self-recovery by reheating it from room temeperature to 1450°C. In addition, the temperature range and the onset of thermal decomposition of Al_2TiO_5 in the temperature range 20 – 1400°C was investigated. The operation conditions of the MRPD were λ = 1.667 Å, 2θ range = 4-138°, step size = 0.1°, counting time ~40-50 s/step, monochromator of 8 Ge crystals (115 reflection), and 32 ^{3}He detectors 4° apart. The relative abundance of phases present was computed using the Rietveld method. The models used to calculate the phase abundance for MRPD were Maslen et al.[19] for alumina, Epicier et al.[20] for Al_2TiO_5, and Howard et al.[21] for rutile. The software used to analyze the data was Rietica 1.7.7.

RESULTS AND DISCUSSION
Effect of Grain Size on Phase Stability
Figure 1 shows the typical neutron diffraction plots of Al_2TiO_5 with CG and MG microstructures before and after thermal decomposition at 1100°C for 10 h to form corundum and rutile. The corresponding diffraction plots for the FG sample are shown in Fig. 2. The good quality of Rietveld refinement plots for three samples are shown in Fig. 3 where "goodness-of-fits" of between 2.5 -3.5 were achieved. The Bragg factors (R_B) obtained for corundum, Al_2TiO_5 and rutile were between 2-5 – 2.7, 3.0 – 3.2, and 1.9 – 2.1 respectively.

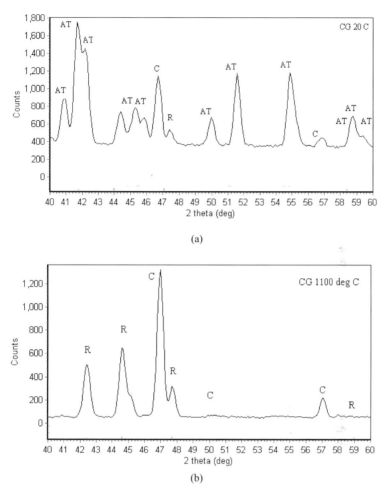

Fig. 1: Typical neutron diffraction plots of as-sintered CG or MG Al_2TiO_5 (a) before and (b) after isothermal decomposition at 1100°C for 10 h. [Legend: AT = Al_2TiO_5; C = corundum; R = rutile]

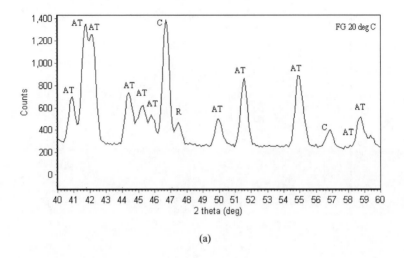

(a)

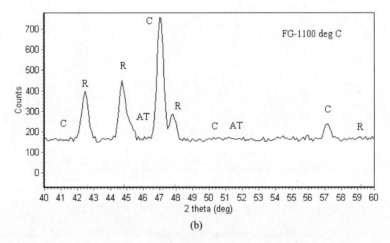

(b)

Fig. 2: Typical neutron diffraction plots of as-sintered FG Al$_2$TiO$_5$ (a) before and (b) after isothermal decomposition at 1100°C for 10 h. [Legend: AT = Al$_2$TiO$_5$; C = corundum; R = rutile]

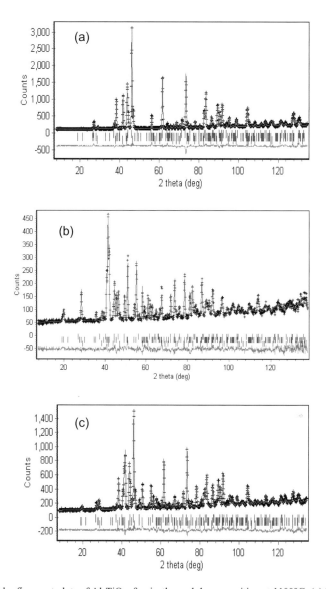

Fig. 3: Rietveld refinement plots of Al_2TiO_5 after isothermal decomposition at 1100°C: (a) FG, (b) MG, and (c) CG.

The profound effect of grain size on the isothermal stability of in air at 1100°C is revealed in Fig. 4. Clearly, coarse-grained Al_2TiO_5 exhibits a slowest rate of thermal decomposition when compared to its medium-grained and fine-grained counterparts. To the best of our knowledge, this is the first time that grain size has been shown to affect the propensity of thermal degradation in Al_2TiO_5. However, it is unclear whether there is a critical grain size associated with this phenomenon. The reason for this grain-size effect is unclear at this stage although it may be closely related to its greater tendency for stress-relief through microcracking as the grain size increases. The microcracking phenomenon is closely related to the material microstructure and thermal expansion anisotropy.[13-15] Below a critical grain size, the elastic energy of the system is insufficient to nucleate microcracks during cooling and thus causing no degradation to the mechanical strength. The density of microcracks increases drastically with concomitant stress-relief once the grain size exceeds a critical value. It is hypothesized that the release of residual stresses helps to reduce phase decomposition.

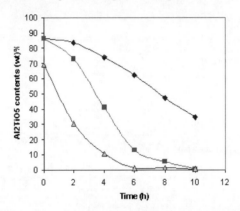

Fig. 4: Effect of Al_2TiO_5 grain size on the propensity of isothermal decomposition at 1100°C in air.
[Legend: CG (♦); MG (■); FG (▲)]

Fig. 5 (a) shows the typical microstructure of as-sintered coarse-grained AT prior to isothermal ageing where the presence of fine microcracks within certain grains is clearly evident. The formation of these microcracks can be attributed to the pronounced thermal expansion anisotropy of AT during cooling from an elevated temperature. The formation of these stress-relief microcracks is believed to impart a low fracture strength but high thermal shock resistance to AT and improved thermal stability. Following isothermal-ageing in air at 1000°C for 10 h, both needle-like and angular particles could be seen to form on the surface of Al_2TiO_5 grains (Fig. 5b). Based on the energy dispersive spectrocopy (EDS) results,[15] these nano-sized particles were identified as surface by-products (ie. Al_2O_3 and TiO_2) of thermally decomposed AT. This may indicate that the initial nucleation process of thermal decomposition of AT is surface-initiated and the growth kinetics are diffusion-controlled or temperature and time dependent.

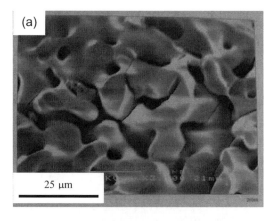

Fig. 5: Scanning electron micrographs of coarse-grained AT (a) before decomposition and (b) after isothermal decomposition at 1100°C. Note the presence of microcracks within certain grains in (a).

The stark contrast in the microstructures of decomposed Al_2TiO_5 with different grain sizes is shown in Figure 6. The light regions in the microstructures indicate the locations of rutile grains following phase decomposition of Al_2TiO_5. A large amount of rutile can be seen in the fine-grained sample which indicates extensive phase decomposition (Fig. 6a). As the microstructure becomes coarser, the degree of phase decomposition appears to become less and is least in the coarse-grained sample (Fig. 6a). This observation is consistent with the neutron diffraction results shown in Fig. 3 above. However, the implication of this work may be a deterrent to materials scientists to develop high-strength nano-structured Al_2TiO_5 which might be highly susceptible to phase decomposition, unless it is stabilized by additives such as MgO, Fe_2O_3 and SiO_2.

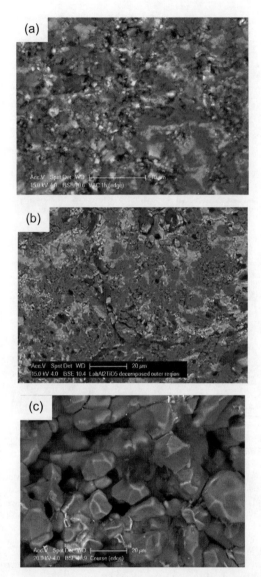

Fig. 6: Scanning electron micrographs showing the microstructures of decomposed Al_2TiO_5 with different grain sizes: (a) FG, (b) MG, and (c) CG. Note the increasing degree of phase decomposition as the grain size decreases from (c) to (a).

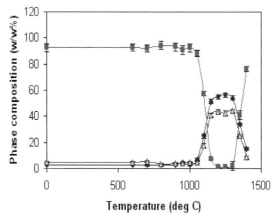

Fig. 7: Thermal stability of Al_2TiO_5 over 20–1400°C as revealed by high-temperature neutron diffraction. Note the display of pronounced thermal decomposition at ~1100 - 1300°C. Errors bars indicate two estimated standard deviations ±2σ. [Legend:■ = Al_2TiO_5; ♦ = Al_2O_3; Δ = TiO_2]

Phenomenon of Self-Recovery

The thermal stability of Al_2TiO_5 in the temperature range 20 – 1400°C as revealed by neutron diffraction is shown in Fig. 7. Clearly, Al_2TiO_5 is stable up to ~1100°C and becomes unstable at between ~1150 - 1300°C. Beyond 1300°C, the thermal decomposition is arrested and the phase stability is restored. This implies that the process of thermal decomposition is reversible or recoverable provided the restricted temperature range of between ~1100 - 1300°C is not transgressed. This process of self-recovery or reversible reaction can be described as follows:

$$Al_2O_3 + TiO_2 \leftrightarrow Al_2TiO_5 \qquad\qquad (1)$$

Figure 8 provides further evidence of self-recovery in decomposed Al_2TiO_5 when it was reheated from room temperature to 1450°C for 2 h. It is clearly shown that self-recovery takes place through the rapid reaction of corundum and rutile to form Al_2TiO_5 with >98 wt% phase purity. This capability of self-recovery further suggests the process of decomposition is spontaneous and reversible as indicated in Equation (1). The implication of this phenomenon is far-reaching whereby it may be possible to restore the decomposed Al_2TiO_5 to its original condition by thermal annealing at >1400°C.

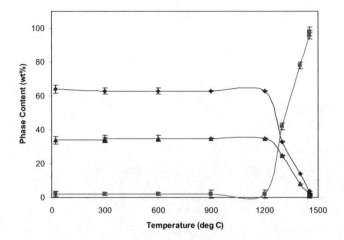

Fig. 8: Reformation of Al_2TiO_5 through self-recovery in decomposed Al_2TiO_5. [Legend: ■ = Al_2TiO_5; ♦ = corundum; ▲ = rutile]

CONCLUSIONS

The effect of grain size on the thermal stability of Al_2TiO_5 at 1100°C and the phenomenon of self-recovery in the temperature range 20-1400°C have been dynamically examined by neutron diffraction. The thermal stability of Al_2TiO_5 increases as the grain size increases probably through the formation of stress-relief microcracks. The process of phase decomposition is reversible and self-recovery occurs readily when decomposed Al_2TiO_5 is re-heated above 1300°C.

ACKNOWLEDGMENTS

This work was supported by funding from the Australian Institute of Nuclear Science and Engineering (AINSE Awards 04/207 & 05/206). We are grateful to our colleague, E/Prof. B. O'Connor, for advice on Rietveld analysis of XRD data. We thank Dr. M. Avdeev of the Bragg Institute of ANSTO for experimental assistance in the collection of MRPD data. We also thank Mr. A. Jones of Alcoa and Ms E. Miller of CMR for assistance with SEM work.

REFERENCES

[1]A.E. Austin and C.M. Schwartz, The Crystal Structure of Aluminium Titanate, *Acta Cryst*. **6**, 812-13 (1953).

[2]B. Morosin and R.W. Lynch, Structure Studies on Al_2TiO_5 at Room Temperature and at 600°C, *Acta Cryst*. B. **28**, 1040-1046 (1972).

[3]H.A.J. Thomas and R. Stevens, Aluminium Ttitanate – a Literature Review. Part 1: Microcracking Phenomena, *Br. Ceram Trans. J*. **88**, 144-90 (1989).

[4]H.A.J. Thomas and R. Stevens, Aluminium Titanate - A literature Review. Part 2: Engineering Properties and Thermal Stability, *Br. Ceram Trans. J.* **88**, 184-190 (1989).

[5]V. Buscaglia, P. Nanni, G. Battilana, G. Aliprandi, and C. Carry, Reaction Sintering of Aluminium Titanate: 1 - Effect of MgO Addition, *J. Eur. Ceram. Soc.* **13**, 411-417 (1994)

[6]G. Tilloca, Thermal Stabilization of Aluminium Titanate and Properties of Aluminium Titanate Solid Solutions, *J. Mater. Sci.* **26**, 2809-2814 (1991).

[7]E., Kato, K. Daimon and Y. Kobayashi, Factors Affecting Decomposition Temperature of β- Al_2TiO_5, *J. Am. Ceram. Soc.* **63**, 355-356 (1980).

[8]R.W. Grimes and J. Pilling, Defect Formation in β-Al_2TiO_5 and its Influence on Structure Stability, *J. Mater. Sci.* **29**, 2245-49 (1994).

[9]M. Ishitsuka, T. Sato, T. Endo and M. Shimada, Synthesis and Thermal Stability of Aluminium Titanate Solid Solutions, *J. Am. Ceram. Soc.* **70**, 69-71 (1987).

[10]B. Freudenberg and A. Mocellin, Aluminum Titanate Formation by Solid-State Reaction of Coarse Al_2O_3 and TiO_2 Powders, *J. Am. Ceram. Soc.* **71**, 22-28 (1988).

[11]B. Freudenberg and A. Mocellin, Aluminium Titanate Formation by Solid State Reaction of Al_2O_3 and TiO_2 Single Crystals, *J. Mater. Sci.* **25**, 3701-3708 (1990).

[12]I.M. Low, D. Lawrence, and R.I. Smith, Factors Controlling the Thermal Stability of Aluminium Titanate in Vacuum, *J. Am. Ceram. Soc.* **88**, 2957-2961 (2005).

[13]I.M. Low, Z. Oo and B. O'Connor, Effect of Atmospheres on the Thermal Stability of Aluminium Titanate, *Physica B: Condensed Matter*, **385-386**, 502-504, (2006).

[14]I.M. Low and Z. Oo, Reformation of Phase Composition in Decomposed Aluminium Titanate, *Mater. Chem. & Phys.* **111**, 9-12 (2008).

[15]A. Jones and I.M. Low, Microstructural Characteristics of Isothermally-Aged Aluminium Titanate Ceramics, pp.185-186 in *Proc. of AUSTCERAM 2002* (Eds. I.M. Low & D.N. Phillips), 30 Sept – 4 Oct. 2002, Perth, WA.

[16]Y. Ohya, Z. Nakagawa and K. Hamano, Crack Healing and Bending Strength of Aluminium Titanate Ceramics at High Temperature, *J. Am. Ceram. Soc.* **71**, C23-C33 (1988).

[17]Y. Ohya and Z. Nakagawa, Grain Boundary Microcracking Due to Thermal Expansion of Al_2TiO_5 Ceramics at High Temperature, *J. Am. Ceram. Soc.* **70**, C184-C186 (1987).

[18]K. Hamano, Y. Ohya and Z. Nakagawa, pp. 129-137 in *Int. Journal of High Tech. Ceram.* Elsevier Science Publishers Ltd., UK. (1985)

[19]E.N. Maslen, V.A. Streltsov, N.R. Streltsova, N. Ishizawa and Y. Satow, Synchrotron X-Ray Study of the Electron Density in α-Al_2O_3, *Acta Crystallographica*, **B49**, 937-980 (1993).

[20]T. Epicier, G. Thomas, H. Wohlfromm and J.S. Moya, High Resolution Electron Microscopy Study of the Cationic Disorder in Al_2TiO_5, *J. Mater. Res.* **6**, 138-145 (1991).

[21]C.J. Howard, T.M. Sabine and F. Dickson, Structural and Thermal Parameters for Rutile and Anatase, *Acta Cryst. B*, **47**, 462-468 (1991).

Nanolaminated Ternary Carbides and Nitrides

TITANIUM AND ALUMINUM BASED COMPOUNDS AS A PRECURSOR FOR SHS OF Ti$_2$AlN

L. Chlubny, J. Lis, M.M. Bućko
University of Science and Technology, Faculty of Materials Science and Ceramics, Department of Technology of Ceramics and Refractories
Al. Mickiewicza 30, 30-059, Cracow, Poland

ABSTRACT
Selfpropagating High-temperature Synthesis (SHS) is very effective method for obtaining numerous materials, including MAX phases such as Ti$_2$AlN, Ti$_3$AlC$_2$, Ti$_2$AlC or Ti$_3$SiC$_2$. These compounds combine features of both ceramic and metallic materials which lead to many interesting potential application. In present work SHS synthesis of Ti$_2$AlN material is presented. Due to the fact that use of elementary powder for SHS synthesis of MAX phases seems to be ineffective, various compounds of titanium and aluminium such as TiAl, Ti$_3$Al, TiN and AlN were used as precursors for synthesis of ternary material. After the synthesis phase composition of obtained materials was examined by XRD method.

INTRODUCTION
Among many covalent materials such as carbides or nitrides there is a group of ternary compounds referred in literature as H-phases, Hägg-phases, Novotny-phases or thermodynamically stable nanolaminates. These compounds have a M$_{n+1}$AX$_n$ stoichiometry, where M is an early transition metal, A is an element of A groups (mostly IIIA or IVA) and X is carbon and/or nitrogen. Heterodesmic structures of these phases are hexagonal, P6$_3$/mmc, and specifically layered. They consist of alternate near close-packed layers of M$_6$X octahedrons with strong covalent bonds and layers of A atoms located at the centre of trigonal prisms. The M$_6$X octahedra, similar to those forming respective binary carbides, are connected one to another by shared edges. Variability of chemical composition of nanolaminates is usually labeled by the symbol describing their stoichiometry, e.g. Ti$_2$AlN represents 211 type phase and Ti$_3$AlC$_2$ – 312 type. Structurally, differences between the respective phases consist of the number of M layers separating the A-layers: in the 211's there are two, whereas in the 321's three M-layers [1-3]. The layered, heterodesmic structure of MAX phases leads to an extraordinary set of properties. These materials combine properties of ceramics, like high stiffness, moderately low coefficient of thermal expansion and excellent thermal and chemical resistance with low hardness, good compressive strength, high fracture toughness, ductile behavior, good electrical and thermal conductivity characteristic for metals. They can be used to produce ceramic armor based on functionally graded materials (FGM) or as a matrix in ceramic-based composites reinforced by covalent phases.
The objective of this work was to examine possibility of obtaining Ti$_2$AlN powders by SHS method with use of various titanium and aluminium based precursors.

PREPARATION
Following the experience gained while synthesising ternary materials such as Ti$_2$AlC, T$_2$AlN and also Ti$_3$AlC$_2$ [4, 5, 6], mostly intermetallic materials in the Ti-Al system were used as precursors for synthesis of Ti$_2$AlN powders. Also other titanium and aluminium based materials such as titanium

nitride (TiN), aluminium nitride (AlN) and elementary powders (Ti and Al) were used as precursors for the synthesis.

Due to relatively low availability of commercial powders of intermetallic materials in the Ti-Al system it was decided to synthesize them by SHS method. At the first stage of the experiment TiAl and Ti_3Al powders were synthesized by SHS method [4, 5, 6]. Titanium hydride powder (TiH_2) and metallic aluminium powder with grain sizes below 10 m were used as sources of titanium and aluminium. The mixture for SHS had a molar ratio of 1:1 and 3:1 respectively (equations 1-2).

$$TiH_2 + Al \rightarrow TiAl + H_2 \qquad (1)$$

$$3TiH_2 + Al \rightarrow Ti_3Al + 3H_2 \qquad (2)$$

These powders were initially mixed using a ball-mill. Then powders were placed in a graphite crucible which was heated in a graphite furnace in the argon atmosphere up to 1200°C. Under these conditions SHS reaction was initiated. After synthesis, products were initially disintegrated in a roll crusher or mortars to the grain size ca. 1 mm and then powders were ground in the rotary-vibratory mill for 8 hours in isopropanol to the grain size ca. 10 m.   As for the other precursors, commercially available powders of aluminium, titanium, aluminium nitride and titanium nitride were used.

The synthesis of more complex ternary compound - Ti_2AlN was conducted by SHS with a local ignition system and with use of various precursors. All sets of precursors for self-propagating high-temperature synthesis (SHS) were set in appropriate stoichiometric ratio and ignited in a nitrogen atmosphere. The expected chemical reactions of equations presented below were synthesis of anticipated compound:

$$4 Ti + 2 Al + N_2 \quad 2 Ti_2AlN \qquad (3)$$

$$2 TiAl + 2 Ti + N_2 \quad 2 Ti_2AlN \qquad (4)$$

$$TiAl + TiN \quad Ti_2AlN \qquad (5)$$

$$TiAl + Ti_3Al + N_2 \quad Ti_2AlN \qquad (6)$$

$$Ti_3Al + Ti + N_2 + Al \quad Ti_2AlN \qquad (7)$$

$$TiN + Ti + Al \quad Ti_2AlN \qquad (8)$$

$$AlN + 2Ti \quad Ti_2AlN \qquad (9)$$

Homogenized mixtures were placed in high-pressure reactor as a loose bed in a graphite holder. The SHS synthesis was initiated by a local ignition and performed at 0.5 MPa of nitrogen in case of reaction were gaseous reagent was necessary.

The X-ray diffraction analysis method was applied to determine phase composition of the synthesized materials. The basis of phase analysis were data from ICCD [7]. Phase quantities were calculated by comparison method, based on knowledge of relative intensity ratios (RIR) [8].

RESULTS AND DISCUSSION

X-ray diffraction analysis proved that TiAl synthesized by SHS was almost pure phase and contained only about 5% of Ti_3Al impurities (Figure 1)[9], while Ti_3Al was a single phase material (Figure 2)[9].

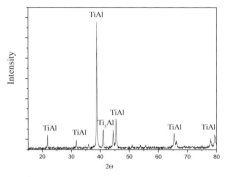

Fig. 1 XRD pattern of the TiAl powders obtained by SHS

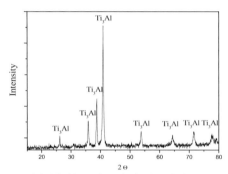

Fig. 2 XRD pattern of the Ti₃Al powders obtained by SHS

The results of XRD analysis of SHS derived Ti_2AlN are presented on Figures 3-9, for reaction 3-9 respectively. As it can be easily observed, the best result was achieved in case of reaction where TiAl and titanium precursors were used (equation 4). The dominant phase was Ti_2AlN accompanied by TiN and Ti_3AlN and Ti_3Al. In case of other synthesis (equations 3, 6, 7, 8, 9) amount of MAX phase varied from ca. 20% to 27% while dominating phase was titanium nitrogen, also many other accompanying phases from Ti-Al-N system were observed. In case of reaction 4 synthesis of MAX phase was not observed and product of reaction consist of titanium aluminium and titanium nitride as it was in substrate mixture but increase of nitride amount was observed, so it can be stated that presence of titanium nitride in reagent mixture prevent formation of MAX phase in favour of formation of regular nitride.

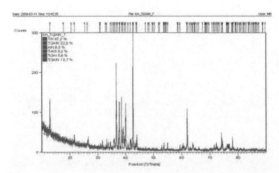

Fig. 3 XRD pattern of product of 4 Ti + 2 Al + N$_2$ synthesis

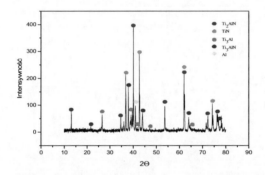

Fig.4 XRD pattern of product of 2 TiAl + 2 Ti + N$_2$ synthesis[9]

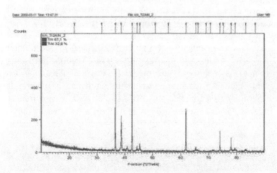

Fig. 5 XRD pattern of product of TiAl + TiN synthesis

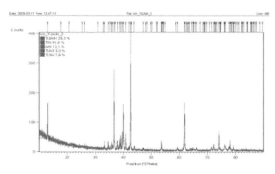

Fig. 6 XRD pattern of product of TiAl + Ti$_3$Al +N$_2$ synthesis

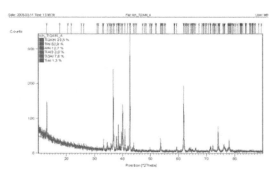

Fig. 7 XRD pattern of product of Ti$_3$Al + Ti +N$_2$ +Al synthesis

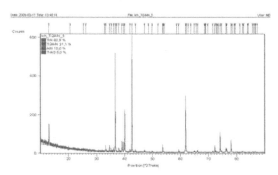

Fig. 8 XRD pattern of product of TiN + Ti +Al synthesis

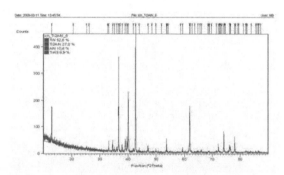

Fig. 9 XRD pattern of product of AlN + 2Ti synthesis

The phase compositions of each product are presented in Table I.

Table I. Phase composition of products obtained by SHS.

Precusor	Composition, % wt.
4 Ti + 2 Al + N$_2$	47% TiN, 22,5% Ti$_2$AlN, 9,5% AlN, 3TiAl$_3$, 4% Ti$_3$Al 14% Ti$_3$AlN
2 TiAl + 2 Ti + N$_2$	57% Ti$_2$AlN, 24% TiN, 11% Ti$_3$Al, 8% Ti$_3$AlN
TiAl + TiN	67% TiN, 33%TiAl
TiAl + Ti$_3$Al +N$_2$	25%Ti$_2$AlN, 52%TiN, 12% AlN, 3% TiAl$_3$, 7% Ti$_3$Al
Ti$_3$Al + Ti +N$_2$ +Al	24% Ti$_2$AlN, 53% TiN, 13% AlN, 2% TiAl$_3$, 8 Ti$_3$Al, 1% TiAl
TiN + Ti +Al	21% Ti$_2$AlN, 61% TiN, 10% AlN, 8% TiAl$_3$
AlN + 2Ti	27% Ti$_2$AlN, 53% TiN, 10% AlN, 10% TiAl$_3$

CONCLUSIONS
    Obtaining of Ti$_2$AlN powders by Selfpropagating High-temperature Synthesis (SHS) with use of titanium and aluminium based precursors is not only possible but also effective and efficient. Low energy consumption and short time of synthesis is the key advantage of this procedure. Intermetallic precursor, namely TiAl seems to be the best as a precursor for synthesizing ternary compound and allows achieving 57% of desired MAX phase in a final product of synthesis. Further elimination of TiN

phase from the product is the main objective of future researches. Basing on previous researches hot-pressing process may be the solution[5]. Powders characterized by highest content of ternary phase were destined to further sintering process and examinations of phase development during sintering and hot pressing process. Mechanical properties of dense, polycrystalline samples will be examined.

Acknowledgements:
This work was supported by the Polish Ministry of Science and Higher Education under the grant no. N 507 2112 33.

5. References
[1]W. Jeitschko, H. Nowotny, F.Benesovsky, Kohlenstoffhaltige ternare Verbindungen (H-Phase). Monatsh. Chem. **94** 672-678, (1963).
[2]H. Nowotny, Structurchemie Einiger Verbindungen der Ubergangsmetalle mit den Elementen C, Si, Ge, Sn. Prog. Solid State Chem. **2** 27, (1970).
[3]M.W. Barsoum: The MN+1AXN Phases a New Class of Solids; Thermodynamically Stable Nanolaminates- Prog Solid St. Chem. **28**, 201-281, (2000).
[4]L. Chlubny, M.M. Bucko, J. Lis "Intermetalics as a precursors in SHS synthesis of the materials in Ti-Al-C-N system" Advances in Science and Technology, **45** 1047-1051, (2006)
[5]L. Chlubny, M.M. Bucko, J. Lis "Phase Evolution and Properties of Ti$_2$AlN Based Materials, Obtained by SHS Method" Mechanical Properties and Processing of Ceramic Binary, Ternary and Composite Systems, Ceramic Engineering and Science Proceedings, Volume **29**, Issue 2, 2008, Jonathan Salem, Greg Hilmas, and William Fahrenholtz, editors; Tatsuki Ohji and Andrew Wereszczak, volume editors, 2008, p 13-20
[6]L. Chlubny, J. Lis, M.M. Bucko M „Preparation of Ti$_3$AlC$_2$ and Ti$_2$AlC Powders by SHS Method", Materials Science and Technology (MS&T) 2009 October 25-29, 2009, Pittsburgh, Pennsylvania, 2205-2213.
[7]"Joint Committee for Powder Diffraction Standards: International Center for Diffraction Data"
[8]F.H.Chung: Quantitative interpretation of X-ray diffraction patterns, I. Matrix-flushing method of quantitative multicomponent analysis. - J.Appl.Cryst., **7**, 513 – 519, (1974a)
[9]L. Chlubny: New materials in Ti-Al-C-N system. - PhD Thesis. AGH-University of Science and Technology, Kraków 2006. (*In Polish*)

INVESTIGATIONS ON THE OXIDATION BEHAVIOR OF MAX-PHASE BASED Ti_2AlC COATINGS ON γ-TiAl

Maik Fröhlich
DLR – German Aerospace Center, Institute of Materials Research
Cologne, Germany

ABSTRACT

The oxidation resistance of the MAX-phase based material Ti_2AlC was investigated as coating on γ-TiAl. A 10 μm thick layer was deposited by low temperature DC magnetron sputtering on the high temperature alloy Ti-45Al-8Nb (at.%). The MAX-phase formation was enhanced after the deposition process by a heat treatment at 1000°C under high vacuum conditions (10^{-6} mbar). The oxidation resistance of the treated coating system was tested at 950°C in air and compared with the results of not annealed samples. In comparison to bare MAX phase material, the exposure tests show no continuous alumina scale formation on top of the annealed coating, similar to the not annealed specimens. The annealed and not annealed coatings failed during cyclic testing after 40 and 60 1h-cycles by spallation. SEM and EDS analysis after 2 1h-cycles showed the growth of a thick mixed oxide scale on top of the annealed MAX-phase coating. Moreover, $α_2$-Ti_3Al and σ-Nb_2Al formed at the substrate/coating interface, indicating Al as well as Ti depletion by enhanced diffusion processes within the Ti_2AlC layer, that presumably leads to fast oxidation.

INTRODUCTION

$M_{n+1}AX_n$ – phases are a group of nanolaminated materials where M is a transition metal, A an A-group element and X either carbon or nitrogen. This class of materials has interesting properties including that of metals (electrical and high thermal conductivity, ductility, thermal shock resistance and machinability) as well as of ceramics (low density, hardness, thermodynamic stability and oxidation resistance)[1-4]. Due to these properties MAX-phases are considered for high temperature applications.

Investigations on the oxidation behaviour of Ti_2AlC revealed an excellent oxidation resistance of that MAX-phase up to 1300°C due to the formation of a continuous protective alumina scale on top[5-9]. Therefore, the group of MAX-phases, especially Ti_2AlC, could be a good candidate for improving the high temperature oxidation resistance of several materials by using it as a coating. Palmquist described different methods to deposit MAX-phase based coatings[2]. They can be successfully synthesized by several chemical (CVD)[2,10] and physical vapour deposition (PVD)[2,11,12] processes such as evaporation or sputtering.

In this work, Ti_2AlC was deposited by magnetron sputtering to increase the high temperature oxidation resistance of γ-TiAl. Engine designers show continued interest in γ-TiAl based titanium aluminides as light-weight structural materials to be used at moderately elevated temperatures of about 700 – 900°C[13-15]. These alloys can meet the requirements for components in automotive combustion engines (exhaust valves and turbocharger wheels) as well as for aero engines (compressor and turbine blades). Especially the low density of approximately 4 g/cm^3 is an attractive property of γ-TiAl based titanium aluminides. But at exposure temperatures above approximately 750°C these alloys show poor oxidation resistance. In this study, Ti_2AlC was applied on γ-TiAl to extend the lifetime of titanium aluminides utilizing the excellent oxidation resistance of the coating material.

EXPERIMENTAL DETAILS

The substrate material used was the γ-TiAl based Ti-45Al-8Nb alloy (in at.%), provided by GfE, Germany. Disc-shaped specimens with 15 mm diameter and 1 mm thickness were machined from extruded rods. Their surfaces were ground using SiC paper up to 4000 grit, polished and finally cleaned.

10 μm thick Ti$_2$AlC coatings were deposited on the substrate material by using the DC magnetron sputter process. The dual source equipment LA 250S (Von Ardenne Anlagentechnik GmbH) allows to pre-treat the sample surface by ion etching as well as to deposit coatings all over the surface by substrate rotation in the center of the vacuum chamber (see Figure 1). The specimens were mounted by wires on the substrate holder. The coatings were deposited by using a Ti$_2$AlC compound target, prepared by Kanthal A.B. The composition of the coatings is presented in Table I.

During the deposition process a temperature of approximately 200°C was measured. For the in-situ synthesis of the Ti$_2$AlC MAX-phase by using DC magnetron sputtering a deposition temperature in the range of 800 – 900°C is necessary[3]. Due to the low process temperature several samples were annealed after coating deposition under high-vacuum conditions (approximately 10^{-6}mbar) at 1000°C for 100h in order to synthesize the Ti$_2$AlC MAX-phase. The parameters for heat treatment based upon previous investigations on Ti-Si based coating systems[16].

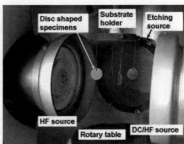

Figure 1. Magnetron sputter equipment LA 250S (Von Ardenne Anlagentechnik GmbH). The Ti$_2$AlC target was mounted on the right sputter source.

After annealing the coatings were investigated by X-ray diffraction analysis (XRD) for phase identification. In comparison, the oxidation resistance of the annealed as well as not annealed coated specimens was studied by performing mass gain measurements during thermal cyclic exposure to air at 950°C. One thermal cycle consisted of 1h heating and 10 min cooling down to approximately 60°C. The samples were examined after 2 cycles as well as after failure to study the evolution of microstructure and phase formation during exposure. For microstructural analysis scanning electron microscopy (SEM) and energy-dispersive X-ray (EDX) spectroscopy were used.

RESULTS AND DISCUSSION

The SEM cross-section of an annealed sample is shown in Figure 2. The annealed coating is approximately 14 μm in thickness. XRD investigations revealed the formation of Ti$_2$AlC as well as TiC phases during annealing (see Figure 3). EDS measurements, presented in Table I, show a higher carbon content compared to the amount of aluminum in the annealed Ti$_2$AlC coating. A higher amount of carbon was deposited during the magnetron sputter process, presumably resulting in the formation of TiC apart from Ti$_2$AlC. Eklund et al. reported about a similar effect in the Ti-Si-C system[17]. Especially carbon and titanium reveal different angular and energy distributions ejecting from the target surface. This finally leads to a higher carbon and lower titanium content in the coating compared to the composition of the compound target.

In comparison to the as coated condition, EDS measurements show an increase of the Ti content and a decrease of the amount of Al after annealing (see Table I). That indicates an outward diffusion of titanium and an inward diffusion of aluminum during heat treatment, presumably resulting in a thicker coating (14 μm) compared to the layer in the as-coated condition with a thickness of 10 μm.

Furthermore, a relatively high amount of niobium (4.2 at.%) was measured in the Ti₂AlC coating (see Table I). Probably, during annealing Nb diffused from the substrate Ti-45Al-8Nb into the coating and partially substituted Ti on the lattice sites. This indicates that a quaternary (Ti,Nb)₂AlC phase was formed by annealing. Obviously, in open literature no information was found about the oxidation resistance of this phase.

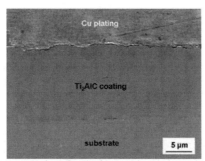

Figure 2. SEM cross-section of a Ti₂AlC coated γ-TiAl sample annealed under high vacuum conditions (10^{-6} mbar) at 1000°C for 100h.

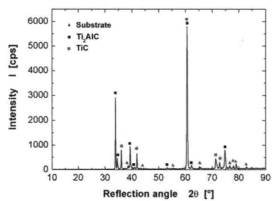

Figure 3. XRD results of a Ti₂AlC coated γ-TiAl sample annealed under high vacuum conditions at 1000°C for 100h; phase identification follows data given by Lin[5] and Wang[8].

Table I. Composition of the Ti₂AlC coating measured by EDS (in at.%).

Condition	Element concentration c [at.%]			
	C	Al	Ti	Nb
As-coated	31.6	28.4	40.0	-
After annealing at 1000°C for 100h under high vacuum conditions (10⁻⁶ mbar)	26.7	23.1	46.0	4.2
After annealing at 1000°C for 100h under high vacuum conditions (10⁻⁶ mbar) and exposure to air at 950°C for 2 1h-cycles	29.7	20.5	46.0	3.8

The oxidation behaviour of the coated material, annealed and not annealed, was tested in air at 950°C under thermo-cyclic conditions. After 2 1h-cycles as well as after failure the samples were removed from the furnace and investigated by SEM and EDS analysis.

After 2 1h-cycles

The not annealed coating was completely oxidized after 2 cycles (see Figure 4a). The formed oxide layer, approximately 12 µm in thickness, consisted predominantly of fast growing TiO_2 apart from small amounts of Al_2O_3. No Nb-oxides could be found. At the substrate/oxide interface a continuous alumina scale was formed. With regard to the fast oxide scale formation the oxygen partial pressure at the substrate/oxide interface could have been too low to form titania. As it can be seen in the Richardson-Ellingham diagram, shown in Figure 5, alumina can be formed at lower oxygen partial pressures in comparison to titania. Therefore, according to the low amount of oxygen at the substrate/oxide interface and the higher Al content in the substrate ($\approx$ 45 at.%) a continuous alumina scale could be formed. That protective scale offers a protection against further fast oxidation for a certain time.

Presumably, the fast oxidation of the not annealed coating was caused by the higher $c_{Ti}/c_{Al}\approx2$ ratio, compared to that of the substrate material: $c_{Ti}/c_{Al}\approx1$. Previous work revealed a comparable oxide scale thickness of approximately 15 µm on bare Ti-45Al-8Nb after 50h at 950°C exposure temperature[18].

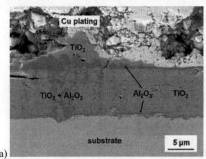 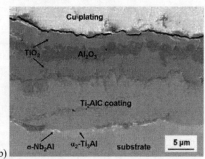

(a)   (b)

Figure 4. SEM micrographs of Ti₂AlC coated samples exposed to air at 950°C for 2 1h-cycles
(a) not annealed
(b) annealed under high vacuum conditions (10⁻⁶ mbar) at 1000°C for 100h.

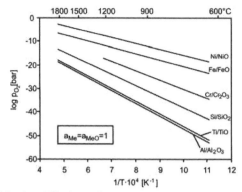

Figure 5. Richardson-Ellingham-Diagramm: Decomposition curves of several oxides; Oxygen partial pressure as a function of temperature[19].

Compared to the not annealed system the MAX-phase based coating on the annealed sample was only partially oxidized (see Figure 4b). A 9 µm thick oxide scale was formed, consisting predominantly of titania. The oxide scale contained of three layers – an outer TiO_2 scale, an intermediate alumina layer and an inner titania scale (see also Figure 6a). The residual coating revealed a thickness of approximately 10 µm, i.e. 4 µm of the Ti_2AlC coating was oxidized. EDS measurements of the residual Ti_2AlC coating show a slightly lower Al content accompanied by a slightly higher amount of carbon compared to the as-treated condition (see Table I).

At the substrate/coating interface the titanium-rich α_2-Ti_3Al as well as the niobium-rich σ-Nb_2Al phases were formed (see Figure 4b and 6b). Presumably, because of Al and Ti consumption by oxide scale formation on top of the coating a concentration gradient of these elements was present. This led to outward diffusion of aluminum and titanium through the $(Ti,Nb)_2AlC$ phase, finally resulting in depletion of these elements at the substrate/coating interface by forming additional intermetallic phases. These phases originated by transformation of the γ-TiAl phase:

$$\gamma\text{-}(Ti,Nb)Al \rightarrow \alpha_2\text{-}(Ti,Nb)_3Al + Al \qquad (1)$$

$$\gamma\text{-}(Ti,Nb)Al \rightarrow \sigma\text{-}Nb_2Al + Ti \qquad (2)$$

The described process can be a possible explanation for the nearly constant composition in the residual MAX-phase coating compared to the as-treated condition (see Table I).

Moreover, the densities of titania (4.2 g/cm³), alumina (4.0 g/cm³) and Ti_2AlC (4.1 g/cm³)[20] are comparable. Therefore, the formation of the 9 µm thick oxide scale on the annealed sample can not be only explained by the oxidation of 4 µm of the MAX-phase based Ti_2AlC coating. Presumably, the outward diffusion of Ti and Al partially led to the increased oxide scale formation, schematically visualized in Figure 7. Aluminum reveals a relatively high diffusion coefficient due to the weak bonding in the Ti_2AlC phase[3,21,22]. Titanium is bonded much stronger. A possible explanation for the increased formation of TiO_2 can be the presence of Nb in the most likely $(Ti,Nb)_2AlC$ phase. Music et al. reported about a bond length of approximately 2.9 Å between niobium and aluminum[23]. The value for the bond length between titanium and aluminum is much higher: 3.9 Å[21]. This indicates that the bonding between Al and Nb is obviously stronger than between Al and Ti. Therefore, it can be assumed that the formation of Al$_2$O$_3$ is decreased by the presence of niobium, finally leading to the increased growth of TiO$_2$. How Nb influences the Ti diffusion in the quaternary (Ti,Nb)$_2$AlC phase is not clear, yet.

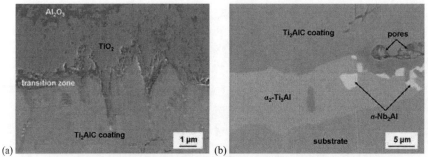

(a)   (b)

Figure 6. SEM micrographs of Ti$_2$AlC coated samples annealed under high vacuum
conditions (10^{-6} mbar) at 1000°C for 100h and tested in air at 950°C for 2 1h-cycles
(a) interface Ti$_2$AlC coating / oxide scale
(b) interface Ti$_2$AlC coating / substrate.

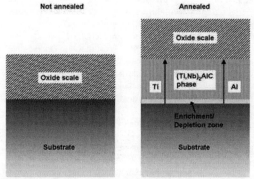

Figure 7. Oxide scale formation on not annealed (left) and annealed (right)
coating systems.

After failure

For further investigations on the development of coating microstructure and oxide scale formation the coated samples were exposed to air at 950°C until failure. The not annealed specimen showed oxide spallation after 60 1h-cycles and the annealed system failed after 40 1h-cycles. These short maximum exposure times indicate low oxidation resistance, especially with regard to the applications and therefore the necessary long-term use of the base material. Contrary to the observations after 2 1h-cycles, the not annealed sample revealed a much thinner total oxide layer thickness of approximately 25 μm compared to that of the specimen annealed under high vacuum conditions (40 μm), shown in Figure 8.

A mixed oxide scale was formed on the surface of the not annealed sample, consisting of titania and alumina with a dense TiO_2 layer on top (see Figure 8a). After 60 1h-cycles a continuous alumina scale was no longer present at the substrate/oxide interface. The additional growth of titania was caused by increase of the oxygen partial pressure by oxygen inward diffusion during further exposure. This finally led to a critical level beyond that titania could grow in addition to alumina. Instead of a protective Al_2O_3 layer a transition zone was formed at the substrate/oxide interface. That zone typically occurs during the oxidation of Nb containing TiAl alloys in air consisting of nitrides (TiN, Ti₂AlN), alumina and σ-Nb₂Al precipitates[24]. The nitride formation is caused by inward diffusion of nitrogen through the thermally grown oxide scale.

The annealed coating system revealed a much thicker oxide scale after only 40 1h-cycles compared to the not annealed sample (see Figure 8b). The outer dense titania and thick alumina layers were more distinct. Furthermore, the microstructure of the inner mixed oxide scale was different. Al_2O_3 was hardly formed there. Moreover, according to studies of Okafor et al. and Chambers et al., EDS measurements revealed the solution of Nb in titania (bright areas in Figure 8b)[25,26]: O: 75.1 at.%; Ti: 17.5 at.%; Nb: 7.4 at.%. The high amount of Nb in the Ti₂AlC coating after annealing, as presented in Table I, led to the formation of high niobium containing titania. Furthermore, a large transition zone at the substrate/oxide interface was observed, characterized by large alumina precipitates and a thick area of α₂-Ti₃Al phase with precipitates of σ-Nb₂Al. The presence of that distinct zone can be explained by the formation of the intermetallic niobium- and titanium-rich phases during oxidation of the Ti₂AlC coating, as described above, enhanced by further growing during further exposure.

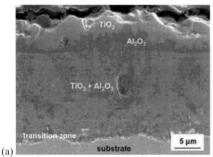 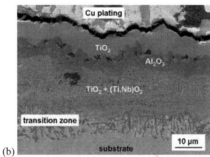

(a)   (b)

Figure 8. SEM micrographs of Ti₂AlC coated samples exposed to air at 950°C
(a) not annealed; failure after 60 1h-cycles
(b) annealed under high vacuum conditions (10^{-6} mbar)
at 1000°C for 100h; failed after 40 1h-cycles.

In comparison, the higher oxidation resistance of the not annealed system was presumably reached because of the formation of the slow growing, continuous alumina scale at the substrate/oxide interface during the first 2 1h-cycles. This protective layer grew slowly until the critical oxygen partial pressure was reached and the fast growing titania was additionally formed.

CONCLUSIONS

According to the excellent oxidation resistance of the Ti₂AlC MAX-phase, this material reveals a high potential as coating material for high temperature applications on titanium aluminides, especially since γ-TiAl and Ti₂AlC contain both, Ti and Al.

Annealing of DC magnetron sputtered Ti$_2$AlC coatings, deposited on Ti-45Al-8Nb (at.%), revealed the formation of a quaternary (Ti,Nb)$_2$AlC phase caused by interdiffusion. This phase formed no protective slow growing Al$_2$O$_3$ scale, leading to early spallation of the thermally grown oxide layer. Outward diffusion of aluminum and titanium caused by consumption at the coating/oxide interface due to oxide formation resulted in the formation of α_2-Ti$_3$Al and σ-Nb$_2$Al intermetallic phases at the substrate/coating interface. This diffusion process was enhanced by weak Al bonding in the Ti$_2$AlC phase as well as, presumably, by the presence of Nb.

The not annealed coating oxidized completely and much faster within the first 2 1h-cycles due to the high Ti content. The fast oxidation resulted in the formation of a continuous protective alumina scale at the substrate/oxide interface, providing acceptable oxidation resistance for a certain time.

Based on the results in this study further investigations will focus on the optimisation of the coating annealing to understand the effect of niobium on the oxidation resistance of Ti$_2$AlC as well as on the diffusion mechanism in the MAX-phase based coating.

REFERENCES

[1]M. W. Barsoum, The M$_{N+1}$AX$_N$ Phases: A New Class of Solids; Thermodynamically Stable Nanolaminates, *Progress in Solid State Chemistry*, **28**, 201-281 (2000).

[2]J.-P. Palmquist, Carbide and MAX-Phase Engineering by Thin Film Synthesis, PhD thesis, University of Uppsala, Sweden (2004).

[3]O. Wilhelmsson, J.-P. Palmquist, E. Lewin, J. Emmerlich, P. Eklund, P. O. Å. Persson, H. Högberg, S. Li, R. Ahuja, O. Eriksson, L. Hultmann, and U. Jansson, Deposition and characterization of ternary thin films within the Ti-Al-C system by DC magnetron sputtering, *Journal of Crystal Growth*, **291**, 290-300 (2006).

[4]M. W. Barsoum, and T. El-Raghy, A Progress Report on Ti$_3$SiC$_2$, Ti$_3$GeC$_2$, and the H-Phases, M$_2$BX, *Journal of Materials Synthesis and Processing*, **5**, 197-216 (1997).

[5]Z. J. Lin, M. J. Zhuo, Y. C. Zhou, M. S. Li, and J. Y. Wang, Microstructural characterization of layered ternary Ti$_2$AlC, *Acta Materialia*, **54**, 1009-1015 (2006).

[6]Z. Li, W. Gao, D. L. Zhang, and Z. H. Cai, Oxidation Behavior of a TiAl-Al$_2$Ti$_4$C$_2$-TiC-Al$_2$O$_3$ in situ Composite, *Oxidation of Metals*, **61**, 339-354 (2004)

[7]R. Ramaseshan, A. Kakitsuji, S. K. Seshadri, N. G. Nair, H. Mabuchi, H. Tsuda, T. Matsui, and K. Morii, Microstructure and some properties of TiAl-Ti$_2$AlC composites produced by reactive processing, *Intermetallics*, **7**, 571-577 (1999).

[8]X. H. Wang, and Y. C. Zhou, High-Temperature Behavior of Ti$_2$AlC in Air, *Oxidation of Metals*, **59**, 303-320 (2003).

[9]J. W. Byeon, J. Liu, M. Hopkins, W. Fischer, N. Garimella, K. B. Park, M. P. Brady, M. Radovic, T. El-Raghy, and Y. H. Sohn, Microstructure and Residual Stress of Alumina Scale Formed on Ti$_2$AlC at High Temperature in Air, *Oxidation of Metals*, **68**, 97-111 (2007).

[10]C. Racault, F. Langlais, R. Naslain, and Y. Kihn, On the chemical vapour deposition of Ti$_3$SiC$_2$ from TiCl$_4$-SiCl$_4$-CH$_4$-H$_2$ gas mixtures, *Journal of Materials Science*, **29**, 3941-3948 (1994).

[11]J. M. Schneider, R. Mertens, and D. Music, Structure of V$_2$AlC studied by theory and experiment, *Journal of Applied Physics*, **99**, 013501 (2006).

[12]W. Garkas, C. Leyens, and A. Flores-Renteria, Synthesis and characterization of Ti$_2$AlC and Ti$_2$AlN MAX phase coatings manufactured in an industrial-size coater, *Advanced Materials Research*, **89-91**, 208-213 (2010).

[13]H. Kestler, and H. Clemens, Production, Processing and Applications of γ-TiAl Based Alloys, in *Titanium and Titanium alloys* (Eds: M. Peters, and C. Leyens), Wiley-VCH Verlag, Weinheim, 351-392 (2002).

[14]H. Baur, and D. B. Wortberg, Titanium Aluminides for Automotive Applications, in *Ti-2003, Science and Technology* (Eds: G. Lütjering, and J. Albrecht), Wiley-VCH Verlag, Weinheim, 3411-3418 (2004).

[15]W. Smarsly, H. Baur, G. Glitz, H. Clemens, T. Khan, and M. Thomas, Titanium Aluminides for Automotive and Gas Turbine Applications in *Structural Intermetallics 2001* (Eds: K. J. Hemker et al.), The Minerals, Metals & Materials Society, Warrendale, 25-34 (2001).

[16]A. Ebach-Stahl, M. Fröhlich, R. Braun, and C. Leyens, Improvement of the High-Temperature Oxidation Resistance of γ-TiAl by Selectively Pre-treated Si-based Coatings, *Advanced Engineering Materials*, **10**, 675-677 (2008).

[17] P. Eklund, M. Beckers, J. Frodelius, H. Högberg, and L. Hultman, Magnetron sputtering of Ti$_3$SiC$_2$ thin films from a compound target, *Journal of Vacuum Science and Technology A*, **25**, 1381-1388 (2007).

[18]M. Fröhlich, Development of High Temperature Coatings on γ-TiAl, PhD thesis, Ernst-Moritz-Arndt University Greifswald, Germany (2008).

[19]C. Leyens, Oxidation and Protection of Titanium Alloys and Titanium Aluminides, in *Titanium and Titanium alloys* (Eds: M. Peters, and C. Leyens), Wiley-VCH Verlag, Weinheim, 187-230 (2002).

[20]W. B. Zhou, B. C. Mei, J. Q. Zhu, and X. L. Hong, Rapid synthesis of Ti$_2$AlC by spark plasma sintering technique, *Materials Letters*, **59**, 131-134 (2005).

[21]M. Magnuson, O. Wilhelmsson, J.-P. Palmquist, U. Jansson, M. Mattesini, S. Li, R. Ahuja, and O. Eriksson, Electronic structure and chemical bonding in Ti$_2$AlC investigated by soft x-ray emission spectroscopy, *Physical Review B*, **74**, 195108 (2006).

[22]J. Wang, Y. Zhou, T. Liao, J. Zhang, and Z. Lin, A first-principles investigation of the phase stability of Ti$_2$AlC with Al vacancies, *Scripta Materialia*, **58**, 227-230 (2008).

[23]D. Music, Z. Sun, A. A. Voevodin, and J. M. Schneider, Ab initio study of basal slip in Nb$_2$AlC, *Journal of Physics: Condensed Matter*, **18**, 4389-4395 (2006).

[24]R. Braun, M. Fröhlich, C. Leyens, and D. Renusch, Oxidation Behaviour of TBC Systems on γ-TiAl Based Alloy Ti-45Al-8Nb, *Oxidation of Metals*, **71**, 295-318 (2009).

[25]I. C. I. Okafor, and R. G. Reddy, The Oxidation Behavior of High-Temperature Aluminides, *JOM*, **51**, 35-40 (1999).

[26]S. A. Chambers, Y. Gao, Y. J. Kim, M. A. Henderson, S. Thevuthasan, S. Wen, and K. L. Merkle, Geometric and electronic structure of epitaxial Nb$_x$Ti$_{1-x}$O$_2$ on TiO$_2$ (110), *Surface Science*, **365**, 625-637 (1996).

ACKNOWLEDGEMENT

The author likes to thank J.-P. Palmquist, Kanthal A.B., for preparing the Ti$_2$AlC target material and A. Ebach-Stahl for technical support with regard to sample annealing.

STUDY OF HIGH-TEMPERATURE THERMAL STABILITY OF MAX PHASES IN VACUUM

I.M. Low[1], W.K. Pang[1], S.J. Kennedy[2], R.I. Smith[3]
[1]Centre for Materials Research, Department of Imaging and Applied Physics, Curtin University of Technology, GPO Box U 1987, Perth WA, Australia
[2]The Bragg Institute, ANSTO, PMB 1, Menai, NSW 2234, Australia
[3]ISIS Facility, Science and Technology Facilities Council, Rutherford Appleton Laboratory, Harwell Science and Innovation Campus, Didcot, Oxfordshire OX11 0QX, UK

ABSTRACT
The susceptibility of two *MAX* phases (Ti_2AlN and Ti_4AlN_3) to high-temperature thermal dissociation in a dynamic environment of high-vacuum has been investigated using *in-situ* neutron diffraction. In high vacuum, these phases decomposed above 1400°C through the sublimation of Ti and Al elements, forming a surface coating of TiN. The kinetics of isothermal phase decomposition was modelled using the Avrami equation and the Avrami exponent (n) of isothermal decomposition of Ti_2AlN and Ti_4AlN_3 was determined to be 0.62 and 0.18 respectively. The characteristics of thermal stability and phase transitions in Ti_2AlN and Ti_4AlN_3 are compared and discussed.

INTRODUCTION

MAX phases are nano-layered ceramics with the general formula $M_{n+1}AX_n$ (n = 1-3), where M is an early transition metal, A is a group A element, and X is either carbon and/or nitrogen. These materials exhibit a unique combination of characters of both ceramics and metals.[1-5] Like ceramics, they have low density, low thermal expansion coefficient, high modulus and high strength, and good high-temperature oxidation resistance. Like metals, they are good electrical and thermal conductors, readily machinable, tolerant to damage, and resistant to thermal shock. The unique combination of these interesting properties enables these ceramics to be a promising candidate material for use in diverse fields, especially in high temperature applications.

However, these *MAX* phases, Ti_3SiC_2 in particular, have poor wear resistance due to low hardness (~4 GPa) and are susceptible to thermal dissociation at ~1400°C in inert environments (e.g., vacuum or argon) to form a protective surface coating of TiC.[6-11] Depth-profiling by x-ray diffraction of Ti_3SiC_2 annealed in vacuum at 1500°C has revealed a graded phase composition with more than 90 wt% TiC on the surface and decreasing rapidly with an increase in depth.[6, 7] A similar phenomenon has also been observed for Ti_3AlC_2 whereby it decomposes in vacuum to form TiC and Ti_2AlC.[10] It follows that this process of thermal dissociation to form protective coatings of binary carbide, nitride or carbo-nitride will also occur in other *MAX* phases such as Cr_2GeC, Ta_4AlC_3, Ti_2AlN, Ti_4AlN_3 and $Ti_2AlC_{0.5}N_{0.5}$. The formation of a graded surface coating such as TiC, TiN or TiCN has the potential to impart high hardness and wear-resistance to the otherwise soft but damage-resistant substrate.[1, 4]

The fundamental knowledge about the thermal stability of technologically important *MAX* phases is still very limited and the actual process of phase dissociation is poorly understood. This limited understanding has generated much controversy concerning the high-temperature thermochemical stability of *MAX* phases.[12-17] We have recently investigated the thermal stability of Ti_3SiC_2, Cr_2AlC, Ti_2AlC and Ti_3AlC_2 in vacuum at up to 1550°C and the results indicated 211 phases to be more resistant to phase dissociation than 312 phases.[8-10] For instance, both Ti_3SiC_2 and Ti_3AlC_2 decompose readily to TiC forming an intermediate phase of Ti_5Si_3C and Ti_2AlC respectively. The apparent activation energies for the decomposition of sintered Ti_3SiC_2, Ti_3AlC_2 and Ti_2AlC were determined to be 179.3, -71.9 and 85.7 kJ mol^{-1}, respectively.[10]

Hitherto, virtually no work has been reported for ternary nitrides such as Ti_2AlN and Ti_4AlN_3. It also remains unknown whether these *MAX* phases will decompose like 211 and 312 ternary carbides via the sublimation of group *M* element and the de-intercalation of group *A* element as follows:

$$M_{n+1}AX_n \rightarrow nMX + A + M \qquad (1)$$

In addition, the 413 phase is expected to form a lower order 211 phase during the initial decomposition process as follows, prior to the sublimation of A and M elements (see equation 1):

$$Ti_4AlN_3 \quad \rightarrow \quad Ti_2AlN + 2TiN \qquad (2)$$

In this paper, we describe the use of high-temperature neutron diffraction to study the dynamic processes of phase stability of Ti_2AlN and Ti_4AlN_3 in high-vacuum. The kinetics of isothermal phase decomposition was modelled using the Avrami equation and the Avrami constants were evaluated. The characteristics of thermal stability and phase transitions in Ti_2AlN and Ti_4AlN_3 are compared and discussed.

EXPERIMENTAL PROCEDURE

Dense hot-pressed cylindrical bars of Ti_2AlN, and Ti_4AlN with diameter 10 mm and height 20 mm were used for the study. The samples were not single phase but contained 6.9 and 0.8 wt% TiN. High temperature time-of-flight neutron diffraction in a vacuum furnace ($\sim 10^{-5}$ torr) fitted with tantalum elements was used to monitor the structural evolution of phase decomposition in the *MAX* phases from 20 to 1800°C in real time. A precision electronic scale with reading to five decimal places was used to measure the weight of samples before and after vacuum-decomposition at a particular temperature. Neutron diffraction data were collected using the Polaris medium resolution, high intensity powder diffractometer at the UK pulsed spallation neutron source ISIS, Rutherford Appleton Laboratory.[18] The diffraction patterns were collected at 20°C and then at between 1500 and 1800°C with a heating rate of 20°C/min. The data acquisition times were 15 min for the room temperature diffraction pattern, and between 15 – 120 min for each of the diffraction patterns collected at elevated temperatures. Normalised data collected in the highest resolution, backscattering detector bank over the *d*-spacing range of ~0.4-3.2Å were analysed using (a) the LAMP software and (b) the Full-Prof software to compute the changes in the phase content of *MAX* phases and *MN* during vacuum-annealing at elevated temperatures. In the former, the integrated peak intensities of lines (013) and (111) were used for calculating the relative phase content of Ti_4AlN_3 and TiN respectively. In the latter, the Rietveld method was used to compute the changes in the phase content of Ti_2AlN and TiN during vacuum-annealing.

The kinetic behaviour of the isothermal decomposition of Ti_4AlN_3 and Ti_2AlN at 1500 and 1550°C respectively was modeled using the Avrami equation to describe the fraction of decomposed *MAX* phase (*y*) as a function of time (*t*) as follows:[19]

$$y(t) = y(0) + \exp(-kt^n) - 1 \qquad (3)$$

where *k* and *n* are time-independent constants for the particular reaction.

RESULTS AND DISCUSSION

(a) Thermal Decomposition and Phase Transitions

The phase evolutions of Ti_2AlN and Ti_4AlN_3 at various temperatures as revealed by in-situ neutron diffraction are shown in Figures 1-3. Samples used in this experiment were not single-phase with TiN as the main impurity. For Ti$_4$AlN$_3$, it began to decompose to TiN quite slowly at 1450°C but became more rapid above 1500°C (Fig. 1). It was almost completely decomposed after annealing at 1600°C for less than 30 min. A total weight loss of ~ 11.6% and 5.9% was observed for decomposition at 1600 and 1500°C respectively, which may be attributed to the release of gaseous Al and Ti by sublimation during the decomposition process. These results concur with our previous work on ternary carbides.[9-12] However, in contrast to Ti$_3$AlC which undergoes an intermediate decomposition to the lower order Ti$_2$AlC,[9-11] such a lower order Ti$_2$AlN was not observed during the decomposition of Ti$_4$AlN$_3$. Nevertheless, such a decomposition process via an intermediate 211 phase cannot be completely ruled out, i.e.

$$Ti_4AlN_3 \rightarrow Ti_2AlN + 2TiN \qquad (4a)$$

$$Ti_2AlN \rightarrow TiN + Al_{(g)} + Ti_{(g)} \qquad (4b)$$

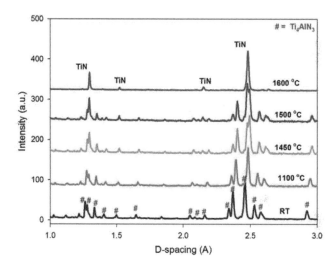

Fig. 1: Phase transitions during the decomposition of Ti$_4$AlN$_3$ at up to 1600°C.

In spite having a higher content of TiN as impurity, Ti$_2$AlN appeared to be much more stable against decomposition than Ti$_4$AlN$_3$. It began to decompose to TiN slowly after 1550°C but became quite rapid at 1600°C (Fig. 2). It was almost completely decomposed after annealing at 1700°C for just over 2 hours and at 1800°C for about 20 min. A total weight loss of more 20.0% was observed for decomposition at 1600°C and above. Below 1600°C, the weight loss was 5.0% and 0.95% at 1550°C and 1500°C respectively.

A closer look at Figure 2 shows that a new phase Ti$_4$AlN$_3$ formed when Ti$_2$AlN was vacuum-annealed at 1600°C and its abundance increased with time and persisted when cooled down to room temperature (Fig. 3). However, it disappeared when the temperature was increased to 1700°C and above. Possible reactions for the formation of Ti_4AlN_3 during the decomposition of Ti_2AlN and its subsequent disappearance are as follows:

$$3Ti_2AlN \rightarrow Ti_4AlN_3 + 2Al_{(g)} + 2Ti_{(g)} \qquad (5a)$$

$$Ti_4AlN_3 \rightarrow Ti_2AlN + 2TiN \qquad (5b)$$

$$Ti_2AlN \rightarrow TiN + Al_{(g)} + Ti_{(g)} \qquad (5c)$$

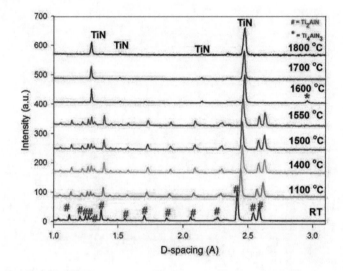

Fig. 2: Phase transitions during the decomposition of Ti_2AlN at up to 1800°C.

As previously mentioned, the weight losses of up to 6% and over 20% in decomposed Ti_2AlN and Ti_4AlN_3 respectively can be attributed to the release of gaseous Al and Ti by sublimation during the decomposition process because the vapor pressures of both Al and Ti exceed the ambient pressure of the furnace (i.e. $\leq 5 \times 10^{-5}$ torr) at ≥ 1500°C.[20] Since the vapor pressure of a substance increases non-linearly with temperature according to the Clausius-Clapeyron relation,[21] the volatility of Al and Ti will increase with any incremental increase in temperature. Figure 4 shows the vapour pressures of various elements at elevated temperature and at a vapour pressure of 5×10^{-5} torr in the vacuum furnace, Al and Ti become volatile at temperature greater than 950 and 1450°C respectively. Thus, at the temperature of well over 1500°C used in this study, both Al and Ti should become volatile and sublime readily and continuously in a dynamic environment of high vacuum. When the vapor pressure becomes sufficient to overcome ambient pressure in the vacuum furnace, bubbles will form inside the bulk of the substance which eventually appear as voids on the surface of decomposed *MAX* phase.[9-12]

The evidence of voids formation can be clearly discerned from the porous surface damage of decomposed Ti_2AlN and Ti_4AlN_3 (Fig. 5). A closer look at Fig. 4 also explains why Ti_3SiC_2 is more resistant to decomposition than Ti_3AlC_2 or Ti_4AlN_3 because Si has a lower vapour pressure than Al. Thus, the use of vapor pressure of elements such as Fig. 4 can be used to predict the susceptibility of *MAX* phases to thermal decomposition. Alternatively, the sublimation pressure of an element can be estimated from the following equation: [22]

$$\ln P^S_{solid} = \ln P^S_{liquid} - \frac{\Delta H_m}{R}\left(\frac{1}{T} - \frac{1}{T_m}\right)$$

(6)

where P^S_{solid} = sublimation pressure of the solid component at the temperature $T < T_m$; P^S_{liquid} = extrapolated vapor pressure of the liquid component at the temperature $T < T_m$; ΔH_m = heat of fusion; R = gas constant; T = sublimation temperature, and T_m = melting point temperature.

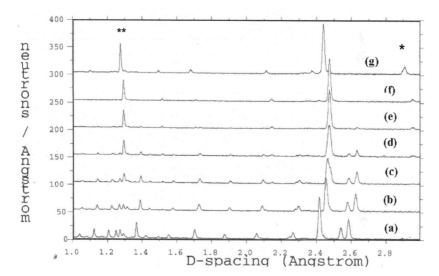

Fig. 3: Diffraction patterns of Ti_2AlN vacuum annealed at: (a) 20°C; (b) 1400°C; (c) 1600°C/10 min; (d) 1600°C/40 min; (e) 1600°C/80 min; (f) 1600°C/130 min, and (g): Cooled to 20°C. Note the formation of Ti_4AlN_3 at 1600°C. [Legend: * = Ti_4AlN_3; ** = TiN]

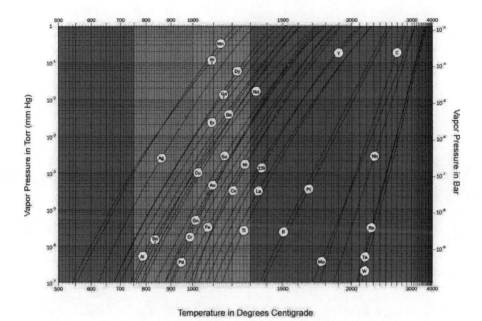

Fig. 4 Vapor pressure of elements at various temperatures.[20]

(i)
[(a) Before decomposition; (b) after decomposition at 1600°C for 440 min.]

(ii)

[(a) Before decomposition; (b-d) after decomposition at 1500°C/120 min, 1550°C/460 min & 1600°C/420 min]

Fig. 5: Surface conditions of (i) Ti_4AlN_3 and (ii) Ti_2AlN before and after thermal decomposition in vacuum.

(b)Isothermal Phase Decomposition and Avrami Kinetics

Fig. 6 shows the isothermal decomposition of Ti_4AlN_3 at 1500°C which was quite rapid initially but slowed down significantly after 30 minutes dwell. More than 40% of Ti_4AlN_3 decomposed after vacuum annealing at 1500°C for 400 minutes. In contrast, the extent of decomposition for Ti_2AlN was much less with only 20% decomposed after 300 minutes dwell at 1550°C (Fig. 7). This implies that Ti_2AlN has a significantly higher resistance to thermal decomposition than Ti_4AlN_3. A possible reason for the high susceptibility of Ti_4AlN_3 to decomposition is either the existence of a lower order 211 phase or the much weaker Al-Ti bonding in this more complex layered compound where Al lies in every fifth layer.[3] In contrast, every third layer in the 211 compound lies the Al atoms, resulting in shorter but stronger Al-Ti bonds which provides more resistance to decomposition via out-diffusion of Al from the bulk to the surface.

During the isothermal decomposition of 211 and 413 phases at 1550 and 1500°C respectively, the Avramii kinetics of decomposition was modeled using Equation (3) and the Avrami constants were evaluated. The Avrami fit of the isothermal decomposition of Ti_4AlN_3 and Ti_2AlN is shown in Fig. 6 and Fig. 7 respectively. The calculated Avrami exponent (n) and Avrami constant (k) for the two *MAX* phases are summarized in Table 1. In general, when the value of *n* is large (e.g. 3 or 4), a 3-dimensional nucleation and growth processes are involved.[23] High values of *n* can also occur when nucleation occurs on specific sites such as grain boundaries or impurities which rapidly saturate soon after the transformation begins. Initially, nucleation may be random and growth unhindered leading to high values for *n*. Once the nucleation sites are consumed the transformation will slow down or cease. Furthermore, if the distribution of nucleation sites is non-random then the growth may be restricted to 1 or 2-dimensions. Site saturation my lead to *n* values of 1, 2 or 3 for surface, edge and point sites, respectively.[19] Since the values of *n* obtained in this study for both 211 and 413 phases are less than 1.0, this implies that the decomposition process is driven by highly restricted out-diffusion of aluminium from the bulk to the surface of the sample and into the vacuum.

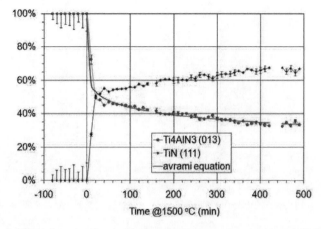

Fig. 6: Time-dependent phase abundance and Avrami fit of isothermal decomposition of Ti₄AlN₃ at 1500°C.

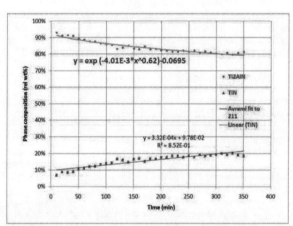

Fig. 7: Time-dependent phase abundance and Avrami fit of isothermal decomposition of Ti₂AlN at 1550°C.

Table 1. Comparison of the decomposition of *MAX* 211 and 413 phases.

MAX phase	Avrami exponent (n)	Avrami constant (k) mol% $\cdot$(min)$^{-n}$
Ti₄AlC₃	0.18	0.37
Ti₂AlN	0.62	4.01×10^{-3}

CONCLUSIONS

The high-temperature thermal stability of Ti_2AlN and Ti_4AlN_3 in a dynamic environment of high-vacuum has been studied using in-situ neutron diffraction. Both 211 and 413 phases were susceptible to decomposition above 1400°C through sublimation of Al and Ti elements, resulting in a surface coating of TiN being formed. The kinetics of isothermal phase decomposition was modelled using the Avrami equation and the Avrami constants (n and K) of isothermal decomposition of Ti_2AlN and Ti_4AlN_3 were determined to be 0.62, 0.18 and 4.01 × 10^{-3}, 0.37 mol% ·(min)$^{-n}$ respectively.

ACKNOWLEDGEMENTS

This work formed part of a much broader project on the thermal stability of ternary carbides which is funded by an ARC Discovery-Project grant (DP0664586) and an ARC Linkage-International grant (LX0774743) for one of us (IML). Neutron beamtime at ISIS (RB920121) was provided by the Science and Technology Facilities Council together with financial support from an AMRFP grant and LIEF grant (LE0882725).

REFERENCES

[1]M.W. Barsoum, The $M_{N+1}AX_N$ Phases: A New Class of Solids: Thermodynamically Stable Nanolaminates, *Prog. Solid State Chem.* **28**, 201-281 (2000).

[2]H.B. Zhang, Y.W. Bao and Y.C. Zhou, Current Status in Layered Ternary Carbide Ti_3SiC_2, A Review, *J. Mater. Sci. Technol.* **25**, 1-38 (2009).

[3]M.W. Barsoum and T. El-Raghy, The MAX Phases: Unique New Carbide and Nitride Materials, *Am. Sci.* **89**, 334-343 (2001).

[4]I.M. Low, Vickers Contact Damage of Micro-Layered Ti_3SiC_2, *J. Europ. Ceram. Soc.* **18**, 709-713 (1998).

[5]I.M. Low, S.K. Lee, M.W. Barsoum and B.R. Lawn, Contact Hertzian Response of Ti_3SiC_2 Ceramics, *J. Am. Ceram. Soc.* **81**, 225-228 (1998).

[6]I.M. Low, Z. Oo and K.E. Prince, Effect of Vacuum Annealing on the Phase Stability of Ti_3SiC_2, *J. Am. Ceram. Soc.* **90**, 2610-14 (2007).

[7]I.M. Low, Depth-Profiling of Phase Composition in a Novel Ti_3SiC_2–TiC System with Graded Interfaces, *Mater. Lett.* **58**, 927-32 (2004).

[8]Z. Oo, I.M. Low and B.H. O'Connor, Dynamic Study of the Thermal Stability of Impure Ti_3SiC_2 in Argon and Air by Neutron Diffraction, *Physica B*, **385-386**, 499-501 (2006).

[9]W.K. Pang, I.M. Low, and Z.M. Sun, In-Situ High-Temperature Diffraction Study of Thermal Dissociation of Ti_3AlC_2 in Vacuum, *J Am. Ceram. Soc.* In press.

[10]W. K. Pang, I.M. Low, B.H. O'Connor, A.J. Studer, V.K. Peterson, Z.M. Sun and J-P Palmquist, Comparison of Thermal Stability in *MAX* 211 and 312 Phases, *J. Physics: Conference Series.* In press.

[11]W. K. Pang, and I.M. Low, Diffraction study of Thermal Dissociation in the Ternary Ti-Al-C System, *J. Aust. Ceram. Soc.* **45**, 39-43 (2009).

[12]W.K. Pang, I.M. Low, B.H. O'Connor, A.J. Studer, V.K. Peterson, Z.M. Sun, and J.-P. Palmquist, Effect of Vacuum Annealing on the Thermal Stability of Ti_3SiC_2/ TiC /$TiSi_2$ Composites. *J. Aust. Ceram. Soc.* **45**, 272-77 (2009).

[13]J.X. Chen, Y.C. Zhou, H.B. Zhang, D.T. Wan, and M.Y. Liu, Thermal Stability of Ti_3AlC_2/Al_2O_3 Composites in High Vacuum, *Mater. Chem. Phys.*, **104,** 109-112 (2007).

[14]X.H. Wang and Y.C. Zhou, Stability and Selective Oxidation of Aluminium in Nano-Laminate Ti_3AlC_2 Upon Heating in Argon, *Chem. Mater.* **15**, 3716-3720 (2003)

[15]M.W. Barsoum and T. El-Raghy, Synthesis and Characterization of Remarkable Ceramic: Ti_3SiC_2, *J. Am. Ceram. Soc.* **79**, 1953-56 (1996).

[16]R. Radakrishnan, J.J. Williams and M. Akinc, Synthesis and High-Temperature Stability of Ti_3SiC_2, *J. Alloys Compd.* **285**, 85-88 (1999).

[17]J. Emmerlich, D. Music, P. Eklund, O. Wilhelmsson, U. Jansson, J.M. Schneider, H. Högberg, and L. Hultman, Thermal Stability of Ti_3SiC_2 Thin Films, *Acta Mater.* **55**, 1479-1488 (2007).

[18]S. Hull, R.I. Smith, W. David, A. Hannon, J. Mayers, and R. Cywinski, The POLARIS powder diffractometer at ISIS, *Physica B*, **180-181**, 1000-1002 (1992).

[19]J.W. Cahn, Transformation Kinetics During Continuous Cooling, *Acta Metallurgica* **4**, 572–575 (1956).

[20]www.veeco.com/library/Learning_Center/Growth_Information/Vapor_Pressure_Data_For_Selected_ Elements/index.aspx

[21]H.B. Callen, *Thermodynamics and an Introduction to Thermostatistics*, published by Wiley, 1985.

[22]B. Moller, J. Rarey, D. Ramjugernath, Estimation of the Vapour Pressure of Non-Electrolyte Organic Compounds via Group Contributions and Group Interactions, *J. Mol. Liq.*, **143**, 52-63 (2008).

[23]A.K. Jena, and M.C. Chaturvedi, *Phase Transformations in Materials.* Prentice Hall. pp. 243-47 (1992).

DETECTION OF AMORPHOUS SILICA IN OXIDIZED MAXTHAL Ti$_3$SiC$_2$ AT 500 - 1000°C

W.K. Pang[1], I.M. Low[1], J.V. Hanna[2], J.P. Palmquist[3]
[1]Centre for Materials Research, Department of Imaging and Applied Physics, Curtin University of Technology, GPO Box U 1987, Perth WA, Australia
[2]Department of Physics, University of Warwick, Gibbet Hill Rd., Coventry CV4 7AL, UK
[3]Kanthal AB, Heating Systems R&D, P.O. Box 502, SE-734 27 Hallstahammar, Sweden

ABSTRACT
This paper describes the use of secondary-ion mass spectrometry (SIMS), nuclear magnetic resonance (NMR) and transmission electron microscopy (TEM) to identify the amorphous silica in Ti$_3$SiC$_2$ oxidised at 500–1000°C. The formation of an amorphous SiO$_2$ layer and its growth in thickness with temperature was monitored using dynamic SIMS. Results of NMR and TEM verify for the first time the direct evidence of amorphous silica formation during the oxidation of Ti$_3$SiC$_2$ at 1000°C.

INTRODUCTION
Ti$_3$SiC$_2$ is a structural ceramic that exhibits the merits of both metals and ceramics.[1-16] This unique combination of properties renders Ti$_3$SiC$_2$ a potential candidate material for high temperature applications. To date, the oxidation properties of Ti$_3$SiC$_2$ have been widely investigated. However, mixed and confusing results have been reported for the oxidation behaviour of Ti$_3$SiC$_2$ in air. For example, the oxidation resistance of Ti$_3$SiC$_2$ was reported[12] to be excellent at temperatures below 1100 C due to the formation of a protective SiO$_2$ surface layer. In addition, although the existence of a protective TiO$_2$ (rutile) has been confirmed by all the researchers,[11-14] the presence of a protective SiO$_2$ film is much more elusive.[4] In addition, the nature and precise composition of the oxide layers formed during oxidation remain controversial, especially in relation to the presence of SiO$_2$ and the graded nature of the oxides formed. Although the existence of crystalline SiO$_2$ during Ti$_3$SiC$_2$ oxidation at temperature above 1200 C has been confirmed,[11-14] the nature of silica formed below 1200 C is still unknown. Based on observations from transmission electron microscopy, scanning electron microscopy and x-ray diffraction, the oxide layer formed was reported to contain a mixture of amorphous SiO$_2$ and crystalline rutile[13-17] at >1100°C. The presence of amorphous silica was also predicted to exist during the oxidation of Ti$_3$SiC$_2$.[18]

In this paper, we describe the use of secondary-ion mass spectrometry (SIMS) to characterise the formation of amorphous silica during oxidation of Ti$_3$SiC$_2$ over the temperature range 500- 1000°C. We verify the existence of the elusive amorphous SiO$_2$ with the aid of transmission electron microscopy (TEM) and solid state [29]Si magic-angle-spinning (MAS) nuclear magnetic resonance (NMR).

EXPERIMENTAL PROCEDURE
Maxthal Ti$_3$SiC$_2$, supplied by Kanthal AB, was used in this study. The compositions of this sample in wt% were 65% Ti$_3$SiC$_2$, 27% TiC and 8% TiSi$_2$. For SIMS analysis, thin slices of 3 mm thick were cut and oxidised in an air-ventilated furnace for 20 min at 500-1000 C. Ring-milled and oxidised Ti$_3$SiC$_2$ powder was used for ^{29}Si NMR analysis.

The near-surface compositions of the oxidized slices were analysed using a Cameca Ims-5f SIMS through the elemental monitoring of Ti, C, Si, and O. A 5.5 keV impact-energy Cs$^+$ ion beam was employed and the beam was scanned across areas of 250 × 250 μm^2. The sputtering times were assumed to be directly proportional to the sputtered depth. With the aid of profoliometry, a constant conversion factor of 7.87 m/s was determined and used to change the sputtering time to the sputtered depth.

High resolution solid state ^{29}Si MAS NMR spectra were acquired at ambient temperatures using an MSL-400 NMR spectrometer (B_o=9.4T) operating at the ^{29}Si frequency of 79.48 MHz. Data of ^{29}Si MAS NMR were acquired using a Bruker 7-mm double-air-bearing probe with single pulse (Bloch decay) methods which utilised high-power ^1H decoupling during data acquisition. The MAS frequencies implemented for these measurements were 5 kHz. All ^{29}Si MAS chemical shifts were externally referenced to tetramethylsilane (TMS) at δ= 0ppm via a high purity sample of kaolinite.

Oxidised slices of Ti_3SiC_2 500-1100°C with thicknesses of 1.5 mm and 0.4 mm were analysed by synchrotron radiation diffraction (SRD) using the reflection-mode (1.5 mm) and transmission-mode (0.4 mm). The reflection-mode SRD experiments were performed on Beamline 20B at the Photon Factory in Japan with energy of 17.7 keV. The transmission-mode experiments were performed on Beamline 2-BM at the Advanced Photon Source (APS) of Argonne National Laboratory with a higher energy of 28.0 keV. TEM was also used to observe the existence of glassy phase (if any) formed in the oxidised powder.

RESULTS AND DISCUSSION
Fig. 1 shows the variation of specific ion yields as a function of sputtering depth for the control sample and the corresponding oxidized Ti_3SiC_2 at various temperatures. As would be expected, the composition within the control sample before oxidation was uniform as indicated by the flat curves of the Ti, Si and C ions. Based on the results reported in the literatures [1-14] that a duplex structure with an outer layer of TiO_2 and an inner mixture layer of SiO_2 and TiO_2 formed during oxidization of Ti_3SiC_2, the crossover point of 28Si and 16O shown in Fig. 1 can be assumed to be the boundary between the TiO_2–rich outer layer and inner mixed layer of SiO_2/TiO_2. The thickness of the measured TiO_2–rich layer as a function of temperature is summarised in Table 1.

Similarly, based on the results of synchrotron radiation diffraction shown in Figs. 2 & 3, the TiO_2 formed at 500°C is anatase and its abundance increases only moderately at up to 800°C before it disappears completely. No crystalline SiO_2 was detected at between 500-1000°C (Fig. 2). However, the existence of a rising background at 2 angles 2 (Fig. 3) indicates the apparent presence of glassy phase in the bulk or *below* the surface when Ti_3SiC_2 was oxidised at 750, 900 and 1100 C.

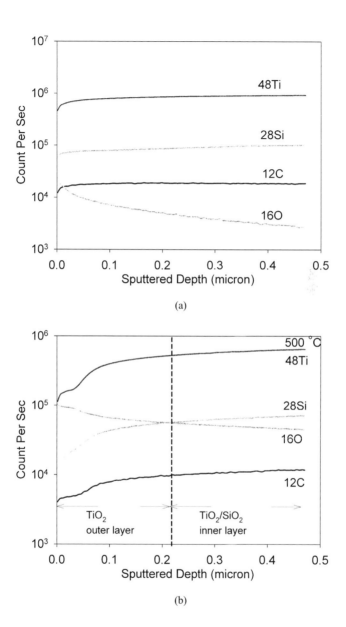

(a)

(b)

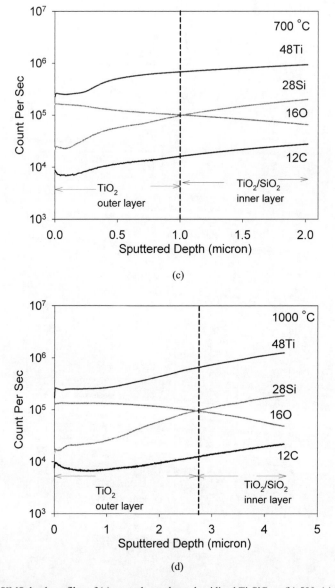

(c)

(d)

Fig. 1: SIMS depth profiles of (a) control sample, and oxidized Ti$_3$SiC$_2$ at (b) 500, (c) 700, and (d) 1000 C.

Table 1: Variation of TiO$_2$–rich layer thickness as a function of temperature.

Temperature (°C)	500	600	700	900	1000
Thickness (µm)	0.21	0.80	1.00	1.67	2.81

From the SIMS results, it is also evident that the intensity of Si increases with an increase in sputtering time or depth which indicates the existence of an inner Si-rich region during the oxidation of Ti$_3$SiC$_2$ (Fig. 1). The Si-rich region implies the existence of inner mixture layer of TiO$_2$ and silica. As previously mentioned, results from synchrotron radiation diffraction showed no presence of crystalline silica in oxidized Ti$_3$SiC$_2$ at 500-1000 C (Fig. 2). This suggests that the silica formed in this temperature range is glassy or amorphous which concurs with the predictions previously made by other researchers [17-18].

In order to verify the claimed existence of amorphous SiO$_2$, ^{29}Si MAS NMR was employed to ascertain the bonding nature of the silica formed during oxidation of Ti$_3$SiC$_2$ at 500, 600, 700, 900, and 1000 C (Fig. 4). As to be expected and in the absence of oxidation, the NMR spectrum (not shown) of as-received Ti$_3$SiC$_2$ showed no silica signal. This is expected because the strongly paramagnetic Ti^{3+} centres will lead to profound broadening resulting in the parent structure beyond observation. In contrast, the ^{29}Si spectra of the oxidised samples show only the semi-condensed/condensed silica phase that has separated from the parent Ti$_3$SiC$_2$ upon oxidation.

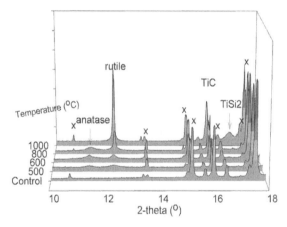

Fig. 2: SRD patterns (reflection-mode) for oxidized Ti$_3$SiC$_2$ at 500-1000 C.

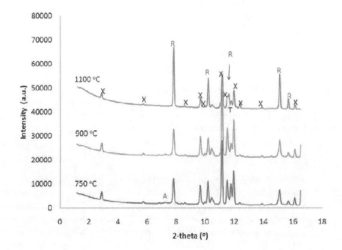

Fig. 3: SRD patterns (transmission-mode) for oxidized Ti₃SiC₂ at between 750 - 1100 C.
[Legend: X = Ti₃SiC₂ phase, T = TiC, A = Anatase, R = Rutile]

It is worth-noting here that the Ti₃SiC₂ [29]Si resonance(s) would appear much closer to 0 ppm (i.e. further downfield) than that observed from the silica phase at ~-112 ppm.[20] The spectrum of Q[2] represents the structures with chains of tetrahedral, Q[3] are sheet silicates and Q[4] are framework silicates. The observed NMR chemical shift peaks are what is expected for amorphous condensed Q[4] silica. It is thus evident that the results of ^{29}Si MAS NMR shown in Fig. 4 provide confirmation of the existence of amorphous silica during the oxidation of Ti₃SiC₂ at 1000 C. The implication of this evidence is wide-ranging for the oxidation of *MAX* phases. It can be postulated that similar amorphous phases (*e.g.* Al₂O₃, GeO₂, SnO₂, *etc.*) are likely to exist during oxidation of other ternary carbides such as Ti₃AlC₂, Ti₃GeC₂, and Ti₂SnC.

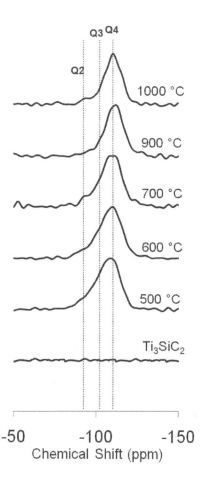

Fig. 4: ^{29}Si MAS NMR spectra of Ti$_3$SiC$_2$ before and after oxidation at 500, 600, 700, 900, and 1000 C.

The confirmation of glassy phase formation during oxidation of Ti$_3$SiC$_2$ at 1000 C can be observed from TEM. Fig. 5a shows the micrograph of a Ti$_3$SiC$_2$ particle oxidized at 600°C. The corresponding diffuse electron diffraction pattern (Fig. 5b) clearly confirms the existence of amorphous phase within the oxidized particle. This suggests that during oxidation of Ti$_3$SiC$_2$, glassy phase commences to form at well below 1000 C and its existence persists even after tridymite or cristobalite has crystallized at 1200 C.[15] This ubiquitous existence of glassy phase during oxidation of Ti$_3$SiC$_2$ is responsible for the formation of an adherent and dense protective surface oxide scale. Lin and co-workers[21] have conducted a comprehensive review on TEM characterization of layered ternary ceramics but they have not observed the formation of any glassy phase in these materials. Similarly, no glassy phase was detected under TEM observations in the oxide scales formed during oxidation of the Ti-Al-C substrates.[22, 23] This implies the importance of Si in the formation of glassy or amorphous SiO_2 phase during oxidation of ternary carbides or nitrides.

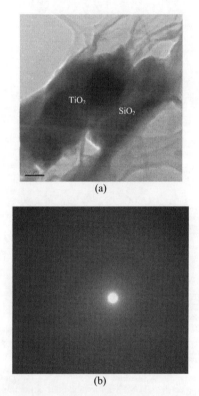

(a)

(b)

Fig.5: (a) TEM micrograph of oxides and (b) its TEM diffraction pattern showing the existence of amorphous silica.

CONCLUSIONS

The composition depth profiles and the growth of oxide layers formed during oxidation of Ti_3SiC_2 in the temperature range 500-1000 C have been characterised by dynamic SIMS. A duplex structure with an outer TiO_2–rich layer and an inner mixed layer of SiO_2/TiO_2 was observed. The existence of amorphous silica in oxidised Ti_3SiC_2 at the temperature range studied has been verified by ^{29}Si-NMR and TEM observations.

ACKNOWLEDGEMENTS

This work formed part of a much broader project on the thermal stability of ternary carbides which is funded by an ARC Discovery-Project grant (DP0664586) and an ARC Linkage-International grant (LX0774743) for one of us (IML). The collection of NMR and SRD data was conducted at ANSTO, Advanced Photon Source and Photon Factory with financial support from AINSE (08/041) and Australian Synchrotron (P21 & P1334). We thank Dr. K.E. Prince and Mr. A. Atanacio of ANSTO for assistance in the collection of SIMS data. We acknowledge the Linkage Infrastructure, Equipment and Facilities Program of the Australian Research Council for financial support (proposal number LE0989759) and the High Energy Accelerator Research Organisation (KEK) in Tsukuba, Japan, for operations support.

REFERENCES

[1] M.W. Barsoum and T. El-Raghy, Synthesis and Characterization of a Remarkable Ceramic: Ti$_3$SiC$_2$, *J. Am. Ceram. Soc.*, **79**, 1953-56 (1996).

[2] E.H. Kisi, J.A.A. Crossley, S. Myhra, and M.W. Barsoum, Structure and Crystal Chemistry of Ti$_3$SiC$_2$, *J. Phys. Chem. Solids*, **59**, 1437-43 (1998).

[3] E. Wu, .H. Kisi, S.J. Kennedy, and A.J. Studer, In-Situ Neutron Powder Diffraction Study of Ti$_3$SiC$_2$ Synthesis, *J. Am. Ceram. Soc.*, **84**, 2281-88 (2001).

[4] N.F. Gao, Y. Miyamoto, and D. Zhang, On Physical and Thermochemical Properties of High-Purity Ti$_3$SiC$_2$, *Mater. Lett.*, **55**, 61-66 (2002).

[5] H. Li, L.M. Peng, M. Gong, J.H. Zhao, L.H. He, and C.Y. Guo, Preparation and Characterization of Ti$_3$SiC$_2$ powders, *Ceram. Int,* **30**, 2289-94 (2004).

[6] J.M. Cardoba, M.J. Sayagues, M.D. Alcala, and F.J. Gotor, Synthesis of Ti$_3$SiC$_2$ Powders: Reaction Mechanism, *J. Am. Ceram. Soc.*, **90**, 825-30 (2007).

[7] R. Radhakrishnan, J.J. Williams, and M. Akinc, Synthesis and High-Temperature Stability of Ti$_3$SiC$_2$, *J. Alloys and Compds.*, **285**, 85-88 (1999).

[8] M. W. Barsoum, The M$_{n+1}$AX$_n$ phases: A New Class of Solids; Thermodynamically Stable Nanolaminates, *Prog. Solid State Chem.*, **28**, 201-81 (2000).

[9] J.-P. Palmquist, S. Li, P.O. A. Persson, J. Emmerlich, O. Wilhelmsson, H. Hogberg, M.I. Katsnelson, B. Johansson, R. Ahuja, O. Eriksson, L. Hultman, and U. Jansson, M$_{n+1}$AX$_n$ Phases in the Ti-Si-C System Studied by Thin-Film Synthesis and Ab-Initio Calculations, *Phy. Rev. B*, **70**, 165401 (2004).

[10] M.W. Barsoum, D. Brodkin, and T. El-Raghy, Layered Machinable Ceramics for High Temperature Applications, *Scripta. Mater.*, **36**, 535-41 (1997).

[11] C. Racault, F. Langlais, and R. Naslain, Solid-State Synthesis and Characterization of the Ternary Phase Ti$_3$SiC$_2$, *J. Mater. Sci.*, **29**, 3384-3392 (1994).

[12] S.B. Li, L.F. Cheng, and L.T. Zhang, The Morphology of Oxides and Oxidation Behaviour of Ti$_3$SiC$_2$-Based Composite at High-Temperature, *Compos. Sci. Tech.,* **63**, 813-19 (2003).

[13] H.B. Zhang, Y.C. Zhou, Y.W. Ban, and J.Y. Wang, Oxidation Behaviour of Bulk Ti$_3$SiC$_2$ at Intermediate Temperature in Dry Air, *J. Mater. Res.*, **21**, 402-408 (2006).

[14] T. Chen, P.M. Green, J.L. Jordan, J.M. Hampikian, and N.N. Thadhani, Oxidation of Ti$_3$SiC$_2$ Composites in Air, *Metall. Mater. Trans. A*, **33**, 1737-42 (2002).

[15] I.M. Low, E. Wren, K.E. Prince, and A. Atanacio, Characterization of Phase Relations and Properties in Air-Oxidised Ti$_3$SiC$_2$, *Mater. Sci. & Eng. A*, **466**, 140-47 (2007).

[16] Z. Oo, I.M. Low, and B.H. O'Connor, Dynamic Study of the Thermal Stability of Impure Ti$_3$SiC$_2$ in Argon and in Air by Neutron Diffraction, *Physica B*, **385**, 449-501 (2006).

[17]M.W. Barsoum, and T. El-Raghy, Oxidation of Ti$_3$SiC$_2$ in Air, *J. Electrochem. Soc.*, **144**, 2508-16 (1997).

[18]T. Okano, T. Yano, and T. Iseki, Synthesis and Mechanical Properties of Ti$_3$SiC$_2$, *Trans. Met. Soc. Jpn.*, **14A**, 597 (1993).

[19]W.K. Pang, I.M. Low, K. E. Prince, and A.J. Atanacio, Mapping of Elemental Composition in Air-Oxidized Ti$_3$SiC$_2$, *J. Aust. Ceram. Soc.*, **44**, 52-55 (2008).

[20]R. Oestrike, W. Yang, R.J. Kirkpatrick, R.L. Hervig, A. Navrotsky, and B. Montez, High-Resolution 23Na, 27Al and 29Si NMR Spectroscopy of Framework Aluminosilicate Glasses, *Geochim. Cosmochim. Acta,* **51**, 2199-2209 (1987).

[21]Z. Lin, M. Li, and Y. Zhou, TEM Investigations of Layered Ternary Ceramics. J. Mater. Sci. Technol., **23**, 145-165 (2007).

[22]Z. Lin, M. Zhuo, Y. Zhou, M. Li, J. Wang, Interfacial microstructure of Ti$_3$AlC$_2$ and Al$_2$O$_3$ oxide scale. Scripta Mater., **54**, 1815-1820 (2006).

[23]Z. Lin, M. Zhuo, Y. Zhou, M. Li, J. Wang, Microstructures and Adhesion of the Oxide Scale Formed on Titanium Aluminum Carbide Substrates, *J. Am. Ceram. Soc.*, **89**, 2964-2966 (2006).

Author Index

Printed and bound by CPI Group (UK) Ltd, Croydon, CR0 4YY